Physics

by Johnnie Dennis

ALPHA

A member of Penguin Group (USA) Inc.

Dedicated to the staff and students of Walla Walla High School and Desales High School in Washington, and to the staff and students of Central High School in Alabama.

ALPHA BOOKS

Published by the Penguin Group

Penguin Group (USA) Inc., 375 Hudson Street, New York, New York 10014, USA

Penguin Group (Canada), 90 Eglinton Avenue East, Suite 700, Toronto, Ontario M4P 2Y3, Canada (a division of Pearson Penguin Canada Inc.)

Penguin Books Ltd, 80 Strand, London WC2R 0RL, England

Penguin Ireland, 25 St. Stephen's Green, Dublin 2, Ireland (a division of Penguin Books Ltd.)

Penguin Group (Australia), 250 Camberwell Road, Camberwell, Victoria 3124, Australia (a division of Pearson Australia Group Pty. Ltd.)

Penguin Books India Pvt. Ltd., 11 Community Centre, Panchsheel Park, New Delhi—110 017, India

Penguin Group (NZ), 67 Apollo Drive, Rosedale, North Shore, Auckland 1311, New Zealand (a division of Pearson New Zealand Ltd.)

Penguin Books (South Africa) (Pty.) Ltd, 24 Sturdee Avenue, Rosebank, Johannesburg 2196, South Africa

Penguin Books Ltd., Registered Offices: 80 Strand, London WC2R 0RL, England

International Standard Book Number: 978-1-59257-691-3
Library of Congress Catalog Card Number: 2007926854

09 08 07   8 7 6 5 4 3 2 1

Interpretation of the printing code: The rightmost number of the first series of numbers is the year of the book's printing; the rightmost number of the second series of numbers is the number of the book's printing. For example, a printing code of 07-1 shows that the first printing occurred in 2007.

Printed in the United States of America

Contents

1 Measuring Quantities in Physics   1
2 Describing Motion Along
  a Straight Line   13
3 Describing Motion in a Plane   23
4 Force and Motion   37
5 The Force of Gravity   49
6 Work and Power   63
7 Mechanical Energy   71
8 States of Matter   81
9 Pressure   93
10 Heat   103
11 Sound   115
12 Atoms, Ions, and Isotopes   131
13 Static Electricity   143
14 Current Electricity   159
15 The Particle Model of Light   175
16 Light and Waves   193

Appendix

Glossary   211
Index   225

Introduction

If you need to review physics for a class or a test, this book provides everything you need to know for an introductory physics class. Therefore, it won't tell you everything there is to know about physics. That's too big of a subject to cover in a little guide like this. However, this book is packed with fun examples and provides a clear and easy-to-follow review of the basics of physics. I hope you find physics and this book as enjoyable as I have.

Extras

Because this book is a refresher and not a tutorial, terms that are defined in the glossary are in *italics*. This will allow you to review definitions as you read along.

Throughout this book you will find sidebars that add bits of information to help you to understand the subject matter being presented.

Physical Harm

Sometimes serves as a warning about possible difficulties that can be avoided and sometimes offers an encouraging word. Some ideas are introduced using this vehicle.

Phun Phacts _____

> Suggestions and tips on how best to master an idea or to remember relationships among concepts. Ideas that are apt to turn up later are emphasized.

Acknowledgments

Thanks to Tom Stevens of Alpha Books for the opportunity to work with this fine publishing house again.

I also want to thank my wife, Shirley, for keeping our home on an even keel while I worked to complete this project. She helped me to keep everything in proper perspective as I dealt with deadlines.

Her careful attention to details of our normal life freed me to concentrate on helping the reader of this book to gain the information she or he needs.

Special Thanks to the Technical Reviewer

The Pocket Idiot's Guide to Physics was reviewed by an expert who double-checked the accuracy of what you'll learn here, to help us ensure that this book gives you everything you need to know about physics. Special thanks are extended to Charles Lynn.

Charles W. Lynn, has a B.S. in Physics from Indiana University and a M.S. in Physics from Miami (OH) University. He is an ASQ certified ASQ Six Sigma Black Belt and Certified Quality Engineer. Charles currently is the Director of Quality and Risk Management for Sekuworks, LLC in Harrison, OH. Charles previous work experience includes positions at Rand McNally, Tyco Healthcare, and Multi-Color Corporation. He currently resides in Maineville, OH with his wife Jennifer and his son Connor.

Trademarks

All terms mentioned in this book that are known to be or are suspected of being trademarks or service marks have been appropriately capitalized. Alpha Books and Penguin Group (USA) Inc. cannot attest to the accuracy of this information. Use of a term in this book should not be regarded as affecting the validity of any trademark or service mark.

Chapter 1

Measuring Quantities in Physics

In This Chapter

- Units of measurement
- Significant figures
- Calculating with rules of scientific notation

Physics is a discipline that requires concentration and critical thinking. A working knowledge of mathematics contributes to the understanding of many concepts of science in general and physics in particular.

This chapter contains information that you will refer to again and again as you read this book. Do not expect to recall right away all of the facts presented here. You do want to become familiar enough with the material to know that you have it here at your fingertips anytime you need it.

Making Measurements

We begin with the systems of measurement used to measure physical quantities. The metric system used includes the MKS and CGS units of measurement. *MKS* is shorthand for *meter, kilogram, and second.* Similarly, *CGS* is a short way of naming *centimeter, gram, and second.* Another system that is used, the British system, is called the FPS system; *FPS* is just a short way of labeling *foot, pound, and second.* Fortunately, the MKS system is so widely used that it is included as part of SI, or International System of Units. We will use all units to practice some mathematics soon. Learn the units of measurement in each of MKS, CGS, and FPS systems. From now on in this book, the three *fundamental quantities* of physics will be measured in these units.

The letters MKS, CGS, and FPS in the titles of those systems are used for three fundamental quantities of physics that we consider early in the study of physics. These quantities are matter (mass is the fundamental measure), space (length is the fundamental measure), and time. I will list them in a table here to help you as we consider these different systems of measurement

Units of Measurement of Physical Quantities

Fundamental Quantity	MKS	CGS	FPS
Time	s	s	s
Space	m	cm	ft
Matter (mass)	kg	g	slug

As you can see, the fundamental quantity of time is measured in the same units in all systems. You will see later that there is a nice mathematical relationship between units in the CGS system and the MKS system. But the FPS system is something else! A slug! Here the slug is the unit of measure of mass in the FPS system. Is it any wonder that students of physics prefer SI units? Unfortunately, the FPS system is the system used widely in the United States even though efforts are still being made to change to SI units.

Recording Measurements and Significant Figures

We should look at a couple of other ideas so that we can use this language to enjoy the game of physics at least on an introductory level. First consider significant figures. A lot of playing physics involves experimentation where measurements are made. First consider *significant figures*. When you make a measurement you should record and report the best observation you can make. In order to do that, you should record every digit that you are sure of, given the measuring instrument that you are using, plus one last digit that you must estimate. That way you and anyone who reads your report will know that any error that may occur will be recorded in that last digit you write down.

Be aware that by recording measurements using the conventions of writing numerals not all written digits are significant. Here is a list of rules that you can follow to make sure your records communicate the same information to anyone who may read your work. Keep this list handy until you know the rules:

1. All nonzero digits are significant. Example: in the number 134 m, there are three significant digits.

2. All zeros occurring between nonzero digits are significant. Example: in the number 104 m, there are three significant digits.

3. All zeros occurring beyond nonzero digits in a decimal fraction are significant. Example: in the number 0.1340 m, there are four significant digits. They are the last four digits.

4. Zeros used to locate the decimal point are not significant. Example: in the number 0.0104 m, there are only three significant digits. They are the last three digits. Also in the number 25,000 mi, there are two significant digits because the three zeros serve only to locate the decimal point.

These rules apply (a) whenever you make a measurement; or (b) when the statement of a problem implies a measurement. Notice that a unit of measurement will be a part of such a record. So whenever you see a numeral with a unit of measurement you will know that significant figures must be observed. That is true with one exception. If the numeral with units is a part of a *defining equation*, all digits are significant. It should also be noted that all numerals without units of measurement contain digits all of which are significant.

You are already familiar with defining equations but you may not have called them by that name before. Consider some of those equations for a moment because using them properly will be of unbelievable value to you later on: 1 hr = 60 min, 1 min = 60 s, 1 mi = 5,280 ft, 1 kg = 1,000 g, 1 m = 100 cm.

All of these are familiar to you, I am sure. Later we will talk about *unit analysis* and at that time I will show you how to use these equations to define helpful disguises of unity. Then you will be able to change from one measurement to another with ease. There are many other defining equations; I will introduce some of them to you and demonstrate how they are used at the proper time. That is an awful lot to remember, so keep your list handy. Some examples of each of these rules will help, so check out the table that follows.

Counting Significant Figures

Recorded Measurement	Number of Significant Figures
102.33 m	5 (by rules 1 and 2)
10.0500 s	6 (by rules 1, 2, and 3)
92,000,000 mi	2 (by rules 1 and 4)

Standardizing Notation

The other notion that I alluded to is *scientific notation*. You probably recall from science classes and newspaper accounts that physics involves some numbers that are very large and some that are very small. That means that we will have to become proficient at writing lots of zeros to represent those numbers unless there is a more efficient method of doing that for us. Scientific notation is just the shorthand we need for writing numbers like three hundred million or six hundred-thousandths.

A number is written in scientific notation in the following way: first remember that the pattern the number in scientific notation must fit is the following $\_.\_ \times 10^n$ where the first blank is for a digit from 1 to 9 inclusive and the second blank is for all remaining significant figures in the original notation multiplied by the power of 10 that makes the number fitting that pattern equivalent to the original number.

For example, the measurement 92,000,000 mi in scientific notation will be 9.2×10^7 mi. The measurement can be read ninety-two million miles and the scientific notation as nine point two times ten million or ninety-two million. Isn't that remarkable? Any doubt about which type of notation is better? If there is, try this number 602,000,000,00 0,000,000,000,000. Would it not be easier to write 6.02×10^{23}? How about a few more examples? We will write these measurements in scientific notation:

Examples of Writing Numbers in Scientific Notation

Original Number	Written in Scientific Notation
250,000 mi	2.5×10^5 mi
0.003 m	3×10^{-3} m
0.005200 g	5.200×10^{-3} g
2.0500 km	2.0500×10^0 km

Phun Phacts

The number 6.02×10^{23} is Avogadro's Number, which refers to the number of things in a mole. Can you imagine a mole of baseballs?

The number 9.2×10^7 mi is the distance from the earth to the sun. A mole is a unit of quantity and is Avogadro's Number of objects. The large numbers listed here serve to illustrate the importance of scientific notation unless you are fond of writing zeros.

Speaking the Language

Now that you have significant figures and scientific notation fresh in your mind, I will review with you methods of calculating some physical quantities using given measurements. In order to keep track of the precision of our calculations, we must abide by some rules listed here (examples in a following section):

- ◆ When adding or subtracting, round off the terms of the sum or difference to the precision of the least precise term in the sum or difference then carry out the operation.

- ◆ When multiplying or dividing, carry out the operation and then round off your answer to have only as many significant figures as the factor in the product or quotient having the least number of significant figures.

In science, in general, and in physics, in particular, we try to tell the best truth possible. You can find a discussion of truth in a lot of sources but probably all will agree with the following observations. In science we do not look for absolute truth—that is, something that is true now, always has been, and always will be true. If you are ever made aware of a scientist saying, "It is absolutely true that this or that scientific event will happen," be watchful because you are about to be scammed! If you want to study absolute truth, consult with your religious advisor or a trusted philosopher.

Even the mathematician does not seek absolute truth. A mathematician's goal is "relative" truth. He assumes certain things to be true absolutely, such as axioms and postulates, but he proves theorems to be true relative to the truth of those ideas he has assumed to be true absolutely. That means that his truth depends upon the quantities he is considering. For instance two plus two is not absolutely four because two drops of oil plus two drops of oil is one drop of oil. Similarly, one plus one is one zero for numbers base two.

The scientist's objective is probable truths. That does not mean maybe a scientific law is true and maybe not. Men and women who make the best measurements possible establish probable truths. They first use their senses to make observations. Then they use their intellect to formulate a hypothesis and to test their hypothesis in the laboratory. After all of that activity, they report the best truth that they can find guided by the theory of probability.

So you can see that we should exert extra effort by observing significant figures and using scientific notation to try to tell the best truth we can as we enjoy the fun of exploring physics. Numbers that are measured are only as good as the instruments used to measure them and the person used to read the measuring instruments. If we record all the numbers that are shown on the display of a calculator after it has multiplied two numbers, we are implying more precision in the instruments used to take the measurements than is true. That is, we would be stating that the calculator is more accurate than the instruments used to determine the measurements. If this is your first exposure to physics or if you are revisiting the difficult subject using this different approach, keep these ideas in mind as you read more and discuss physics.

Calculations Using Significant Figures

Now that I have given you a reason why we use certain ideas, I will show you how to calculate some quantities using significant figures to point out how that notion helps us to tell the best truth possible. Not absolute truth, but probable truth; also since you know that mathematics is the language of physics, you should know that we are borrowing some ideas from mathematicians that involve relative truth.

Suppose you want to add and subtract some measurements of distance. Arrange the sum or difference so that the decimal points are in a vertical line as in the following examples: the first column for each operation below is the original set of terms to be combined. The second column in each case is the set of terms rounded to the precision of the least-precise term in the sum or difference. When rounding numbers use the following rules:

- ◆ If the number being rounded off is greater than five, then drop the number and increase the preceding number by one.

- ◆ If the number being rounded off is less than five, then drop the number and leave the preceding number unchanged.

- ◆ If the number being rounded off is exactly five and the number preceding is odd, then drop the five and increase the preceding number by one. If the number preceding the five is even, then drop the five and leave the preceding number unchanged. This rule helps you to round up as many numbers as you leave unchanged.

Add these measurements:

283.6 cm	283.6 cm
34.621 cm	34.6 cm
91.25 cm	91.2 cm
8.36 cm	8.4 cm
	417.8 cm

Subtract these measurements:

478.348 m	478.3 m
332.1 m	332.1 m
	146.2 m

Now to multiply and divide use the same rounding procedures but follow the rules for calculating using scientific notation.

Multiply these measurements:

$$151 \text{ ft}$$
$$\underline{46 \text{ ft}}$$
$$906$$
$$\underline{604}$$
$$6,900 \text{ ft}^2$$

Divide these measurements:

$$\frac{480.6\,\text{m}^2}{47.8\,\text{m}} = 10.1 \text{ m}$$

Notice that the units of measurement behave just like numerals in calculations, that is, ft × ft = ft^2 and m^2/m = m. You will get a lot of practice with this later, but I wanted to focus your attention on this important property.

The Least You Need to Know

◆ Identify the fundamental units of measurement in the MKS, CGS, and FPS systems.

◆ Three fundamental quantities of physics are space (length is fundamental measure), matter (mass is the fundamental measure), and time.

◆ Significant figures are those digits an experimenter records that she is sure of plus one very last digit that is doubtful.

◆ Make calculations using scientific notation, which are those involving numbers written in the form _._ × 10^n.

Chapter 2

Describing Motion Along a Straight Line

In This Chapter

- Motion (speed and velocity)
- Calculating velocity
- Uniform motion

A body in motion has speed. The distance a body travels per unit of time is its velocity. And for a body traveling in a straight line, its speed and velocity are the same. This chapter deals with the motion of a body in a straight line traveling in one direction.

Speed

The amount of speed a *moving* body has on a straight line depends on how long it takes to make a change in position—that is, the amount of time required to go from an initial position to a final position on the line.

Motion

The following symbols are used to represent some concepts:

- x_0 = initial position, or position when we begin measuring time
- x_f = final position, or position at the end of measured time
- v_0 = initial speed, or velocity when we begin measuring time
- v_f = final velocity, or speed at the end of measured time
- $\overline{v}$ = average velocity, or speed over the time interval
- t_0 = initial time, or the reading on a stopwatch to begin measuring time
- t_f = final time, or the reading on a stopwatch at the end of the measured time interval
- Δ = symbol used to represent change

We use x to represent the position at any time. The change in position will be $\Delta x = x_f - x_0$. The change in time can be expressed as $\Delta t = t_f - t_0$. Similarly, if v represents the velocity at any time, then $v_f - v_0 = \Delta v$ represents the change in velocity. The average velocity is the change in position divided by the change in time: in symbols, $\overline{v} = \dfrac{\Delta x}{\Delta t}$, which is $\overline{v} = \dfrac{x_f - x_0}{t_f - t_0}$.

Distances x can be marked off along the line from an initial position to the final position. Suppose we measure all distances relative to the initial position so that $x_0 = 0$ and $x_f = x$. Furthermore, suppose that we start the stopwatch when the object first moves from the initial position and stop it when it reaches its final position so that $t_0 = 0$ and $t_f = t$. That means that $\overline{v} = \dfrac{x_f - x_0}{t_f - t_0} = \dfrac{x - 0}{t - 0} = \dfrac{x}{t}$.

In other words, the *average velocity* is equal to the length of the line segment joining the initial position and the final position, x, divided by the time, t, required for the object to travel that length.

Therefore, from $\overline{v} = \dfrac{x}{t}$ the expression $x = \overline{v}t$ represents the position at any time assuming that the object is always traveling with the same motion.

Describing Uniform Motion

Speed is not always the same throughout the trip of an object. It can be, for example, zero at times. Average speed treats speed as if an object were in uniform motion, motion that is characterized by a constant velocity or speed. Whenever a situation in physics states or implies uniform motion, the speed is constant. If the speed or velocity is constant, the object has uniform motion.

Except for the fact that we are now talking about constant speed, everything said about motion in the last section is true of uniform motion. Here is a list of the symbols used to describe uniform motion:

- v = the constant velocity
- x = the distance traveled
- t = the time the object is in motion

Uniform motion is described simply as $x = vt$. Since the velocity is constant, the distance traveled is found by calculating the product of the average velocity and the time. The average velocity is the same as the velocity at any time because the velocity is constant.

Suppose you drive along the interstate, set your cruise on $80 \frac{km}{hr}$, and continue on your way for 30 min. How far will you travel? You probably know that 40 km is the correct answer. You are given a constant speed of $80 \frac{km}{hr}$ and a time of 30 min., both having one significant figure. That means that your car is traveling with uniform motion and since $x = vt$, $80 \frac{km}{hr} \times \frac{1}{2} hr$.

The emphasis here is not on significant figures and scientific notation. We have reviewed those ideas and need only to practice to become proficient using them. We now consider the units of measure and the units of our final answer. Using the information from the last section, here is the detailed analysis of the units. Since $x = vt$, $x = 80 \frac{km}{hr} \times 30$ min. The question is, "How do we get km?" We were given that the constant speed is $80 \frac{km}{hr}$ and the time is 30 min and we know that we want distance in km in this problem. By using the defining equation 60 min = 1 hour, we can define unity to be disguised as $\frac{hr}{min}$.

That is, dividing both sides of that defining equation by 60 min we get $1 = \dfrac{1\,\text{hr}}{60\,\text{min}}$.

We can multiply any quantity by 1 without changing the value of the quantity. We just change its appearance. Therefore,

$80\,\dfrac{\text{km}}{\text{hr}} \times \dfrac{1\,\text{hr}}{60\,\text{min}} \times 30\,\text{min} = \dfrac{80\,\text{km}}{60\,\text{min}} \times 30\,\text{min} = 40\,\text{km}$

because $\dfrac{30\,\text{min}}{60\,\text{min}} = \dfrac{1}{2}$.

Express 6.0×10^1 mi/hr in ft/s. Use the following defining equations: 5,280 ft = 1 mi, 60 s = 1 min, 60 min = 1 hr. Disguise unity (1) using each of those equations and write the result: $6.0 \times 10^1\,\dfrac{\text{mi}}{\text{hr}} \times$ $5{,}280\,\dfrac{\text{ft}}{\text{mi}} \times \dfrac{1\,\text{hr}}{60\,\text{min}} \times \dfrac{1\,\text{min}}{60\,\text{s}}$, that is, $6.0 \times 10^1\,\dfrac{\text{mi}}{\text{hr}} \times$ $1 \times 1 \times 1$, which is still $6.0 \times 10^1\,\dfrac{\text{mi}}{\text{hr}}$, but because of the different disguises of unity (1), the final answer will be $88\,\dfrac{\text{ft}}{\text{s}}$.

Express 1 year in seconds. Use the defining equations: 1 yr = 365 days, 1 day = 24 hr, 1 hr = 60 min, and 1 min = 60 s. Then disguise unity from each of those equations and write the following: 1 yr $\times$ $365\,\dfrac{\text{da}}{\text{yr}} \times 24\,\dfrac{\text{hr}}{\text{da}} \times 60\,\dfrac{\text{min}}{\text{hr}} \times 60\,\dfrac{\text{s}}{\text{min}}$, which is another way of writing 1 yr $\times 1 \times 1 \times 1$ (which is still 1 yr) but the disguises of unity result in an answer of 3.1536×10^7 sec, the number of seconds in a year.

Remember that all of the digits are significant in defining equations and 1 yr was not measured.

Understanding Acceleration

In physics speak, if a car speeds up, the car is said to accelerate. If it slows down, it is said to have negative acceleration or it decelerates. Acceleration is expressed in terms of velocity.

The speeding up or slowing down implies a change in velocity. We need to identify some symbols necessary to discuss acceleration:

- a = acceleration
- v_f = final velocity or the velocity at the end of the time interval
- v_0 = initial velocity or velocity at the beginning of the time interval
- t_0 = initial time or time at the beginning of the interval
- t_f = final time or time at the end of the interval

Using this information we can find Δv and Δt then define acceleration as follows: $a = \dfrac{\Delta v}{\Delta t} = \dfrac{v_f - v_0}{t_f - t_0}$, that is, acceleration is the change in velocity divided by the change in time or the rate of change of velocity.

Since $a = \dfrac{\Delta v}{\Delta t}$, Δv will have m/s as units and Δt is measured in s. Now $\dfrac{\Delta v}{\Delta t}$ will have units of $\dfrac{\frac{m}{s}}{s} = \dfrac{m}{s} \times \dfrac{1}{s} = \dfrac{m}{s^2}$ and is read meters per square second or meters per second per second.

It is high time to expand the table of units of measurement.

Units of Measurement of Physical Quantities

Derived Quantities	MKS	CGS	FPS
Speed	m/s	cm/s	ft/s
Acceleration	m/s^2	cm/s^2	ft/s^2

Describing Uniformly Accelerated Motion

Recall that we were able to describe the motion of an object by using its average velocity. We will use that same idea here but we must keep in mind that we want to describe the motion of an object that has a constant acceleration. That makes this a very different discussion. We use the same symbols as we have used before to describe motion. There are three basic ideas you need to develop the solution to any problem concerning an object that has constant acceleration. Some of the algebraic solutions you will want to be able to construct are outlined here as well.

Any time that you consider a problem that states or implies an object moves with constant acceleration these three ideas apply:

- $x = \overline{v}t$ Distance is equal to the average velocity times the time.

◆ $\overline{v} = \dfrac{v_f + v_0}{2}$ Definition of average velocity
where v_f and v_0 are given or implied. v_f is
the velocity at the end of the interval of time
and v_0 is the velocity at the beginning of the
time interval.

◆ $a = \dfrac{v_f - v_0}{t}$ The definition of acceleration
where t is the length of the time interval.

Look at this example. You are in a car that was
initially at rest but is now traveling with acceleration a and continues to accelerate until it reaches a
final speed of v_f. How far did you travel? Use the
procedure for solving a physics problem mentioned
earlier. Solve the problem first in general (algebra-
ically), and then substitute given information for
this particular case. In general $x = \overline{v}t$, that is, distance equals average velocity times the time. Use
the definition of $\overline{v}$ to find that $x = \left(\dfrac{v_f + v_0}{2} \right) t$.

Using the definition of acceleration we find that
$x = \left(\dfrac{v_f + v_0}{2} \right) \left(\dfrac{v_f - v_0}{a} \right) = \left(\dfrac{v_f^2 - v_0^2}{2a} \right)$ for the general solution. The particular solution is obtained by
substituting the information implied in the statement of the problem, $v_0 = 0$, and finding that
$x = \dfrac{v_f^2}{2a}$. We still have a solution in algebraic symbols but the beauty of this solution is that you have
solved all problems of this type.

Remember that this is a discussion of motion in one direction along a straight line. Velocity and speed are closely related and may be used interchangeably under these conditions. You describe uniform motion by the simple statement $x = vt$ and uniformly accelerated motion is described using three ideas: $x = \bar{v}t$, $a = \dfrac{v_f - v_0}{t}$, and $\bar{v} = \dfrac{v_f + v_0}{2}$.

The Least You Need to Know

- The average velocity in a straight line is $\bar{v} = \dfrac{v_f + v_0}{2}$.

- Uniform motion in terms of distance, speed, and time is $x = vt$.

- Uniformly accelerated motion may be described using these three equations $a = \dfrac{v_f - v_0}{t_f - t_0}$, $x = \bar{v}t$, and $\bar{v} = \dfrac{v_f + v_0}{2}$.

Chapter 3

Describing Motion in a Plane

In This Chapter

- ◆ Vectors and scalars
- ◆ Vector algebra
- ◆ Uniform circular motion

This chapter will focus on the need for both vector and scalar quantities and distinguish between them.

Understanding Vectors and Scalars

A vector is a quantity that (1) has magnitude, (2) has direction, and (3) obeys a law of combination (the law of combination is the commutative law of addition).

A scalar is a quantity that has magnitude only. The magnitude includes the units of measurement, of course. Some examples are distance, speed, and time, to mention just a few. We discussed all of these in Chapter 1.

The numeral 3 is the symbol used to represent an abstract idea we refer to as Three. The numeral 3 is unique in that it represents one and only one number. An arrow of given length and direction represents a unique vector quantity.

Recognizing and Dealing With Vectors

In Figure 3.1, you see a representation of two vectors. Included also is a label for each. The symbols $\vec{A}$ and $\vec{B}$ are used to refer to the arrows representing the corresponding vectors so that the vectors may be discussed without drawing arrows. In terms of the new symbols, we say $\vec{A} = \vec{B}$.

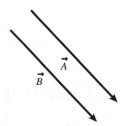

Figure 3.1

Vectors $\vec{A}$ and $\vec{B}$ demonstrate equal vectors. $\vec{A}$ and $\vec{B}$ are algebraic symbols used to name the arrows that represent vectors.

Vector Addition, a New Way of Adding

Now that it has been established what is meant by equal vectors, consider the law of combination. Look at the two vectors in Figure 3.2.

Obviously, vectors $\vec{D}$ and $\vec{C}$ are not equal. What is true about adding these two vectors is true for adding any two vectors, even equal vectors.

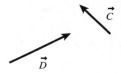

Figure 3.2

Vectors $\vec{D}$ and $\vec{C}$ are unequal vectors.

Follow these rules to add the vectors in Figure 3.2: (1) construct vector $\vec{C}$ at some convenient place in the plane, (2) construct vector $\vec{D}$ with its foot exactly at the tip of the head of vector $\vec{C}$, (3) draw a new vector from the foot of $\vec{C}$ to the head of $\vec{D}$. Call the new vector $\vec{R}$, which is called the sum of $\vec{D}$ and $\vec{C}$. Using the new algebra of vectors that being developed, it can be stated for this case $\vec{C} + \vec{D} = \vec{R}$.

Figure 3.3

The vector sum demonstrates a method of adding vectors.

In Figure 3.4, instead of starting with $\vec{C}$, begin with $\vec{D}$ and add the same two vectors. That is, construct first $\vec{D}$ and then place the foot of $\vec{C}$ exactly at the head of $\vec{D}$. Draw the new vector from the foot of $\vec{D}$ to the head of $\vec{C}$ and compare it with the vector $\vec{R}$ that we found in the previous sum.

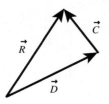

Figure 3.4

The vector sum demonstrates the law of combination.

Refer to Figure 3.5 for an example of combining more than two vectors.

Three different vectors are used in this part of the discussion. The order $\vec{B} + \vec{C} + \vec{A}$ was chosen, but any order for the terms may be used and the result is the same for the given vectors.

Neither the magnitude nor the direction of any of the vectors in Figure 3.5 was changed when they were added together. When adding vectors, be sure to preserve magnitude and direction of every vector used when calculating a sum of vectors.

Use north as the direction vertically upward toward the top of the page and use south for the direction vertically downward. East will be the direction to the right of the page and west the direction to the left. If the direction angle is between any two of these four, measure the acute angle between the vector and a horizontal line you draw through the foot of the vector. If the angle is 30° upward from the horizontal, the direction will be either East30° North or West30° North. That is, the direction is upward and to the right of the page or upward to the left of the page.

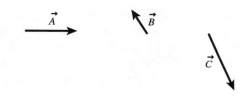

Figure 3.5

The vector sum $\vec{B} + \vec{C} + \vec{A}$ *generalizes the combination rule.*

For your information, the vectors in Figure 3.5 were measured using a ruler and a protractor. Here is a record of those measurements: $\vec{A}$ = 1.9 cm E, $\vec{B}$ = 1.1 cm W55°N, and $\vec{C}$ = 2.4 cm E67°S. Add these vectors together using the rules stated in this section along with the aid of the example for adding three vectors. Notice 1.9 + 1.1 + 2.4 no longer equals 5.4 in the addition of vectors.

A New Subtraction Also

Suppose you want to find the difference in two vectors, $\vec{A}$ and $\vec{B}$. The operation of subtraction can be done in one of two ways: either $\vec{A} - \vec{B}$ or $\vec{B} - \vec{A}$. The rules for subtraction are (1) write down which difference you want to calculate, (2) construct the vectors so that they have their feet together—that is, their feet are at the same point, and (3) draw a vector from the head of the second to the head of the first. The notation used in step one establishes identification of the first vector and the second vector. The first one listed in that notation is the first, and the second one constructed is the second.

Look at Figure 3.6, where two vectors are given and an example of subtraction worked out in detail. As a check, look to see if vector $\vec{B}$ plus the new vector $\vec{A} - \vec{B}$ would add to the original vector $\vec{A}$.

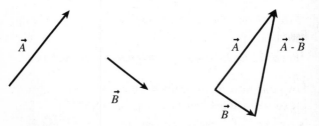

Figure 3.6

Subtracting vectors is an illustration of the rules for subtraction.

Complications of Multiplication in Vector Algebra

There are several types of vector multiplication but only *scalar multiplication*, which is the process of multiplying a vector by a scalar, is discussed here.

Suppose a vector $\vec{A}$ is given and the vector $4\vec{A}$ is to be calculated. Knowing how to add vectors and previous experience suggests that $\vec{A} + \vec{A} + \vec{A} + \vec{A} = 4\vec{A}$. In other words, scalar multiplication is just a quick way to add vectors. That means that n $\vec{A}$ includes $^1\!/_4\, \vec{A}$ (where $n = {}^1\!/_4$) as a vector with the same direction as $\vec{A}$ but only $^1\!/_4$ the length of $\vec{A}$. Multiplying by $^1\!/_4$ is the same as dividing by four so that means that a vector can be divided by a scalar to produce a new vector so long as the scalar is not 0. That leads to another idea. Suppose $n = -1$. Calculating $(-1)\,\vec{A}$ or $-\vec{A}$ yields a vector with the same length as $\vec{A}$ but a direction that is opposite that of $\vec{A}$.

The new vector $-\vec{A}$ is a vector with the same length as $\vec{A}$ but a direction that is opposite that of $\vec{A}$. It behaves just like scalar opposites. That is, -3 is the opposite of 3, and $3 + (-3) = 0$. So $(-\vec{A}) + \vec{A} = \vec{0}$.

The vector algebra is expanded to include scalar multiplication, legal scalar division, the negative of a vector, and also a 0 vector. Now that the negative of a vector has been defined, the negative of either vector in Figure 3.6 can be determined and added to the other vector.

Is Breaking Up Hard to Do?

Think of an inverse process of adding vectors. Suppose you start with a vector, a resultant could be its name, and the problem to be solved is that of finding the parts that were combined originally to make up the given vector.

The process of identifying the parts of a given vector is called *resolution*. The parts are called the *components* of the vector. While the first solution is a good one and not too difficult, looking for and naming specific components can be a challenge. Resolving a vector into specific components can be very helpful in solving problems.

This same procedure can be used to find two mutually perpendicular components of any vector. Given a vector $\vec{A}$, construct two mutually perpendicular lines through the foot of $\vec{A}$. Construct the projection of the vector on each of those lines. Place an arrowhead on each of the components at the points on *l* corresponding to the projections on the tip of the arrow representing $\vec{A}$.

Among other things, these components make it possible to calculate the magnitude of $\vec{A}$ by using the Pythagorean theorem. Figure 3.7 will give you a detailed look at the resolution of $\vec{A}$ into two components that are mutually perpendicular.

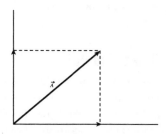

Figure 3.7

The resolution of $\vec{A}$ into two mutually perpendicular components illustrates the identification of two crucial vector components.

Position and Displacement

Think back to the discussion of motion along a straight line and use vectors to discuss motion in a plane as well. To begin with, describe the object moving along the straight line. Let its initial position be the reference point. Its initial position is $\vec{X}_0$ and its final position is $\vec{X}_f$ both of which are vector quantities. When it moved from one end of the line segment to the other, it had a change in position of $\vec{\Delta X} = \vec{X}_f - \vec{X}_0$.

Notice that this is the difference in two vectors and that means that $\vec{\Delta X}$ is a vector drawn from the head of position $\vec{X}_0$ to the head of position $\vec{X}_f$.

Understanding Velocity

Take the displacement of the straight line motion and divide by the time required to make the displacement to see what is meant by this.

$\dfrac{\vec{\Delta X}}{\Delta t} = \vec{V}$ is the equation resulting from that operation, and in words it states that the displacement divided by the time is equal to the average velocity. Velocity is a vector quantity, as you can see. It results from the division of a vector $\vec{\Delta X}$ by the scalar Δt or, in this case, t the time for the displacement to take place. The velocity vector has the same direction as $\vec{\Delta X}$ and has speed for its magnitude.

If it is said that the car is traveling $60\dfrac{\text{mi}}{\text{hr}}\text{N}$, then it cannot be read from the speedometer, and it is not speed. The velocity of the car in that case is $60\dfrac{\text{mi}}{\text{hr}}\text{N}$.

The velocity would be defined by a combination of a reading from the speedometer and a reading from a compass.

Understanding Acceleration

Consider a situation where a car is driven toward an intersection at 20.0 km/hr E when it makes a sharp turn along a circular arc at the intersection without changing speed and is now traveling at 20.0 km/hr N.

If the velocity changed, then acceleration is implied. In Figure 3.8, you see a vector diagram of the final velocity and the initial velocity along with the calculated change in velocity. The change in velocity took a small amount of time t, or Δt if you prefer.

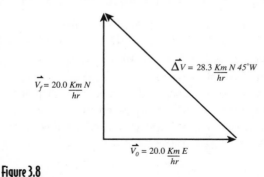

Figure 3.8

The calculated change in velocity implies acceleration.

The change in velocity $\overrightarrow{\Delta V}$ that took place in time $\Delta t = 3.00$ s, for instance, implies that there is acceleration $\overrightarrow{a} = \dfrac{\overrightarrow{\Delta V}}{\Delta t}$. Calculate that quantity from the diagram and along with the assumed time see that there definitely is vector acceleration

$$\overrightarrow{a} = \frac{28.3 \text{ km / hr N45°W}}{3.00 \text{ s}} \times \frac{1 \text{ hr}}{60 \text{ min}} \times \frac{1 \text{ min}}{60 \text{ s}} \times \frac{10^3 \text{ m}}{\text{km}} =$$

$2.62 \dfrac{\text{m}}{\text{s}^2}$ N45°W. The noteworthy thing to realize here is that this is an object with a constant speed and yet it accelerates!

Uniform Circular Motion

Almost everyone has whirled some object about her head at a constant speed. A string tied to the object and the other end of the string held in the hand defined the circular path of the object. She probably even released the object at some point and observed it to fly away in a straight line that is tangent to the circular path at the instant of release.

Speed and Velocity in a Circular Path

There is one and only one vector acceleration that results from the motion of the object in a *circular path* at a constant speed. The speed can be calculated in terms of the time required for the object to make one complete revolution or circular path.

Given that the speed is constant, the time required to make one complete revolution is called the *period of motion*.

The speed is calculated to be $v = \dfrac{2\pi R}{T}$, where R is the radius of the circle and T is the period of the motion.

Let $\vec{V}$ represent velocity and v represent speed. A position vector $\vec{R}$, length R, that is drawn from the center of the circle along a radius to a point on the circle locates the position of the object on the circular path relative to the center of the circle. Two positions of an object moving in a circular path at a constant speed have been identified in Figure 3.9. The two positions can be anywhere on the path.

$\vec{R}_0$ represents the position at the beginning of a time interval Δt or t, either may be used to represent a small time interval. $\vec{R}_f$ is the position of the object at the end of the time interval. The corresponding velocities of the object are represented by vectors tangent to the path at those positions.

Centripetal Acceleration

Suppose that the speed of an object is given to be v. That means that $v = v_0 = v_f$ because the speed is constant. Look at the diagram in Figure 3.9.

Even though the speed is constant, the velocity changes at every instant. Construct the velocity vectors with their feet together so that $\vec{V}_f - \vec{V}_0 = \Delta \vec{V}$ can be calculated. Remember that the magnitude of $\vec{V}_f$ equals the magnitude of $\vec{V}_0$ equals v but the magnitude of $\Delta \vec{V}$ is ΔV, which the length of the change in velocity vector determined by the geometry of the triangle in the construction.

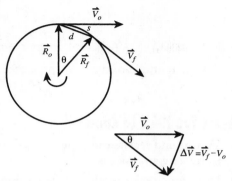

Figure 3.9

The object with uniform circular motion is located by position vectors $\vec{R}_0$ and $\vec{R}_f$.

The instantaneous acceleration is a vector quantity. It has the same direction as the change in velocity vector as it gets close to zero. The magnitude of this acceleration is calculated to be equal to the square of the speed divided by the radius of the circle. The direction of the change in velocity vector, and the acceleration, would then be directly opposite the direction of the position vector, always pointing toward the center of the circular path. The acceleration for uniform circular motion is called *centripetal acceleration*.

Express centripetal acceleration completely by $\vec{a}_c = \dfrac{v^2}{R}$ (toward the center of the circle). If the speed is not given, the speed can be expressed in terms of the period then find $\vec{a}_c = \dfrac{\left(\dfrac{2\pi R}{T}\right)^2}{R} = \dfrac{4\pi^2 R}{T^2}$ which is the magnitude of the centripetal acceleration.

The Least You Need to Know

- ◆ Vectors and scalars can explain the difference in velocity and speed.
- ◆ Uniform circular motion is the motion of an object in a *circular path* at a constant speed.
- ◆ Acceleration is defined by the equation $\vec{a} = \dfrac{\Delta \vec{V}}{\Delta t}$.

4

Force and Motion

In This Chapter

- ◆ Graphical analysis
- ◆ A practical application of a vector
- ◆ Components of force

Kinematics is the study of the description of motion, and until now, that's been the focus of this book. Dynamics is the study of the causes of motion, and that's what we'll look at in this chapter.

It is well known that if an object moves, it must either be pushed or pulled. The push or pull applied to the object is called a force. If the object is given a push it tends to move away and if it receives a pull it tends to move toward the source of the pull. That means that force is a vector quantity. Force is a derived quantity so the units of force in each of the systems of measurement must be identified.

Mathematical Models and Physical Relationships

It is challenging to make the connection between science and mathematics especially algebra. One purpose of this section is to make that connection. In order to gain as much as possible from information that is shared here, it is recommended that the reader review ideas involving graphical analysis in a mathematical reference book.

Applying Graphical Techniques to Experimental Data

We know that if we apply a force to an object that is free to move it will accelerate. An experiment was done using five arbitrary unit forces and five arbitrary unit masses and the acceleration was measured under circumstances given in the following table. The acceleration is the reacting variable and the force is the managed variable.

Acceleration Versus Force (Mass Is Held Constant of 1 Unit)

Acceleration (m/s^2)	Force (Arbitrary Unit)
0.20	1.0
0.40	2.0
0.60	3.0
0.80	4.0
1.0	5.0

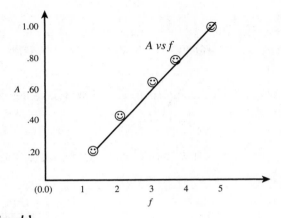

Figure 4.1

The graph of A vs. f suggests a direct variation.

The graph of the data in Figure 4.1 looks like a straight line that would pass through the origin. Using the ideas just reviewed, it is logical to suggest that $A \propto f$ when the mass of the object is held constant. That means that the acceleration increases as the force applied to the object increases. That information is sufficient at this stage because there is a third variable, m, involved.

The next part of the experiment was done by making the mass, m, the managed variable and the acceleration, A, the reacting variable. The acceleration is again measured in m/s² and the mass in arbitrary units. The table is a record of the data.

The force applied is one unit in each trial. The graph in Figure 4.2 looks like a power relation where the exponent is a negative integer. Graphing techniques are used to check that possibility by beginning with $n = -1$. The graph in Figure 4.2 suggests that there is an inverse variation so the logical place to begin the search for the correct power is with the case $n = -1$. That is, it appears that $A \propto \dfrac{1}{m}$ and graphing this information will help us to decide whether that is the relationship we are looking for.

Acceleration Versus Mass (Force Held Constant of 1 Unit)

Acceleration (m/s²)	Mass (Arbitrary Unit)
0.800	1.0
0.400	2.0
0.270	3.0
0.200	4.0
0.160	5.0

The graph of acceleration versus mass with a constant force of one unit suggests an inverse variation.

Before doing that, though, remember that there are three variables involved.

Considering that result we may think if $A \propto \dfrac{1}{m}$ then $\dfrac{1}{A} \propto m$; that is, as m increases then A decreases, by the first statement.

Use the second statement for your graph because you can introduce the third variable by substituting 1 for f since f is one unit throughout this part of the experiment. That means you can plot $\dfrac{1}{A}$ vs m on the next graph, and if a straight line results, then $A \propto \dfrac{1}{m}$ and hence $\dfrac{f}{A} \propto m$.

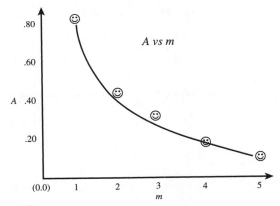

Figure 4.2

A vs. M with a constant force suggests an inverse variation.

The graph in Figure 4.3 shows that $\dfrac{f}{A} \propto m$ and that means that $f \propto mA$ and $f = kmA$.

The constant of proportionality is k, which is usually calculated by finding the slope of the straight line that was discovered experimentally. Since f and m are arbitrary units the slope of the curve is not very meaningful. Take that last expression and use the MKS system to give the constant a meaning by using units that are not arbitrary. In the expression $f = kmA$, use the MKS system $f = (k)(\text{kg})(\text{m/s}^2)$—that is, mass is measured in kg and acceleration is measured in m/s^2.

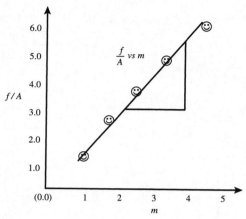

Figure 4.3
The graph of $\frac{f}{A}$ vs. m shows that $\frac{f}{A} \propto m$.

The assignment of units for k was made by saying

let $k = \dfrac{N}{\frac{\text{kg} \cdot \text{m}}{\text{s}^2}}$ so that when you substitute k into

the equation you get $f = \left(\dfrac{N}{\frac{\text{kg} \cdot \text{m}}{\text{s}^2}} \right)\left(\dfrac{\text{kg} \cdot \text{m}}{\text{s}^2} \right) =$

$N = 1$ newton. The assignment was made in honor of Sir Isaac Newton because of his work with force along with many other things, including his invention of the calculus.

The expression now becomes $f = ma$ because $k = 1$ and the force is measured in newtons or N in the MKS system—that is, $1 N = \dfrac{1\,\text{kg} \cdot \text{m}}{\text{s}^2}$. Similarly, in the CGS system 1 dyne $= \dfrac{1\text{g} \cdot \text{cm}}{\text{s}^2}$ and in the FPS system $1 \text{ pound} = \dfrac{1 \text{ slug} \cdot \text{ft}}{\text{s}^2}$. The pound is abbreviated lb and there is no short way of writing dyne.

The Vector Nature of Force

The expression $F = ma$, that was developed from the experiment, is called Newton's second law. We can identify force as a vector quantity in that equation by writing $\vec{F} = m \, \vec{a}$ where $\vec{F}$ is a vector, m is a scalar, and $\vec{a}$ is a vector. The equation states that if a single force is applied to an object that is free to move, the object will accelerate in the direction of the applied force with a magnitude that is directly proportional to the magnitude of the force. When used with units of measurement, the hyphen (-) and dot (.) means multiplication.

If several forces are acting on the same point in a body, we may add them together two at a time by drawing a diagram that begins with their feet together. Such a diagram is called a free-body diagram. Apply that idea in the solution of this problem.

A sports-utility vehicle is stuck on a muddy road. Two ropes are tied to the front bumper and a man pulls on one of the ropes with a 2.50×10^2 N force that makes an angle of 30.0° with a line down the middle of the road. Another man pulls on the other rope with a force of 3.00×10^2 N at an angle of 50.0° with that same line. What is the total force exerted on the vehicle? Both ropes are pulled parallel to the ground. Refer to Figure 4.4, which is a free-body diagram of the object being pulled and the forces applied.

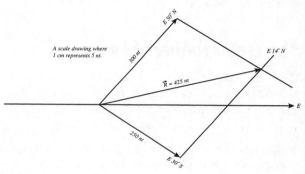

A scale drawing where 1 cm represents 5 nt.

E 50° N

300 nt

E 14° N

$\vec{R}$ = 425 nt

E

250 nt

E 30° S

Figure 4.4

The vector diagram of forces illustrates the parallelogram method of adding two vectors.

The method of adding the forces in Figure 4.4 works only for the addition of two vectors at a time. The method of adding vectors earlier, called the closed polygon method, works for any number of vectors. The method used here is called the parallelogram method. The answer is $\vec{F}$ = 423 N at an angle of 14.4° with the line down the middle of the road. The calculated answer differs from the *graphic solution* by about one half of one percent. Notice that, by Newton's second law, the vehicle will tend to accelerate at an angle of a little more than 14° with the middle of the road if these are the only two forces acting on the vehicle.

Components of Forces

Suppose we must apply 352 N of force to the handle of a lawn mower, and the handle makes an angle of 49° with the horizontal. How much of the force pushes parallel to the surface of the ground? Refer to Figure 4.5 for a graphic solution.

The graphic solution shows that only 235 N of the applied force acts to push the lawn mower parallel to the ground. Notice that we resolved the applied force into two mutually perpendicular components. Remember it was stated earlier that these two components would have important practical applications, and this is one.

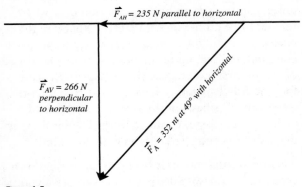

$\vec{F}_{AH}$ = 235 N parallel to horizontal

$\vec{F}_{AV}$ = 266 N perpendicular to horizontal

$\vec{F}_A$ = 352 nt at 49° with horizontal

Figure 4.5

The diagram of the components of a force illustrates the resolution of a force into mutually perpendicular components.

Net Force and Uniform Motion

The forces discussed so far have included component forces and total or resultant forces. It is understood that the single resultant force can be used to replace many forces acting on an object. Newton's second law shows what a single force tends to do to an object. Taking all of these ideas into account, the question might be what happens if we consider all forces acting on an object and arrive at one single force that can replace all of the forces acting on the object? That force is what we call the net force. The net force on an object can give us an indication of the type of motion an object experiences.

The implication for the motion of a body is that an object can travel with uniform motion only when the net force on the object is zero. Otherwise the object would follow Newton's second law and accelerate in the direction of the net force. The so-called Newton's first law enables you to explain the case for a net force of zero. Newton's first law of motion actually had its basis in the ideal experiments of Galileo in the seventeenth century. The ideal experiments were conducted only in Galileo's mind.

That means that the only way you can travel down the interstate at 80 miles per hour is to drive your car in such a way that the net force on your car is zero. It also means that when the car is brought to a sudden stop, the driver tends to continue traveling at a constant speed in the direction the car was traveling initially. All of these notions imply that there are forces other than the ones we have encountered so far in this book.

Newton's third law states that for every action there is an equal and opposite reaction. That means when we push on the wall with our hands, the wall pushes back on our hands with a force equal in magnitude but opposite in direction to the force applied by our hands. The important thing to note here is that there are two bodies involved in the application of this law. Our hand provides the action and the wall must react.

Suppose that we know there are only three forces in the same plane acting on an object. The forces are $F_1 = 20.0$ dynes E15°N, $F_2 = 30.0$ dynes W15°N, and $F_3 = 40.0$ dynes S15°W. What is the net force on the object? Refer to Figure 4.6 that is a graphic solution of this problem. The map convention is used to indicate direction. $\vec{F}_1$ and $\vec{F}_2$ were added using the parallelogram method, then that resultant was added to $\vec{F}_3$ to get the net force, $\vec{F}_1 + \vec{F}_2 + \vec{F}_3$. Figure 4.6 shows the final result to be 37.0 dynes W54.0°S for the net force. The graphic solution is sufficiently accurate for the purposes of solving vector problems in this book.

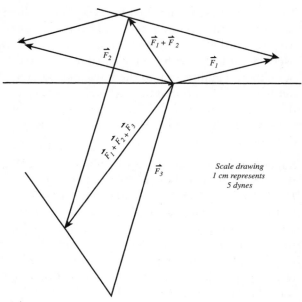

Figure 4.6

A graphic solution for the net force due to three different forces acting at a point uses the parallelogram method of adding vectors.

The Least You Need to Know

+ Discuss the use of graphical analysis to identify the relationship between force and acceleration.

+ Identify and use the units of force in each of the systems of measurement.

+ Recognize the application of Newton's laws of motion to describe the motion of an object.

Chapter 5

The Force of Gravity

In This Chapter

- Gravitational force
- Weight or mass
- Influence on projectile motion
- Motion of planets

This chapter deals with that ever present force of gravity that is largely responsible for the sagging bags under aging eyes and holds us on or near the earth's surface.

A Force at a Distance

Notice that there is nothing attached to your body like the forces discussed in earlier chapters. Gravity is a force similar to several other forces that act at a distance.

Atmosphere and Air Friction

Any object that is allowed to fall near the surface of the earth experiences the same force. It is the force of *air friction*. Figure 5.1 helps to visualize the two forces acting on an object, with mass m, falling near the surface of the earth.

Figure 5.1

The motion of an object can be described using two forces on a falling object.

The object in Figure 5.1 appears to have a net force acting downward. When you calculate the net force you find $\vec{F}_{net} = \vec{F}_A + \vec{F}_g$. This is a vector sum that you can read as follows: the net force is equal to the sum of the force of air friction ($\vec{F}_A$) and the force of gravity ($\vec{F}_g$). That means that $\vec{F}_A = 0$, or at least so small it is insignificant. You can make $\vec{F}_A$ as close to zero as you like by choosing a high-density object (say 2 cm to 5 cm diameter steel ball) with a small surface area. You can see that the net force on such an object is $\vec{F}_{net} \approx \vec{F}_g$. Note that this is very special. The object is said to be a *freely falling* body in this case.

Terminal Velocity

Terminal velocity is the maximum velocity of a falling object in air. The air friction on a falling body will increase with velocity until the upward drag of air friction is equal in magnitude to the downward force of gravity on the body. At that point there is no net downward force on the body, and its downward speed no longer changes.

Acceleration Due to a Force Acting at a Distance

Since $\vec{F}_{net} = \vec{F}_g$, then $m\,\vec{a}_{net} = m\,\vec{a}_g$. Because the object is dense and does not gain or lose mass as it falls, $\vec{a}_{net} = \vec{a}_g$. The acceleration of gravity is so important it is assigned a special symbol of its own. The acceleration of gravity is represented as g with a direction radially inward toward the center of the earth or a direction downward as you would describe it locally.

Phun Phacts

Remember that the terms velocity and speed are interchangeable only when motion is along a straight line in one direction.

Describing the Motion of a Falling Object

If you assume ideal conditions where the object is not influenced by air friction as it falls, then the vertical motion is uniformly accelerated motion. The acceleration is the acceleration of gravity that is constant near the earth's surface. When a problem that involves an object falling near the surface of the earth is read, it is understood that the acceleration is $-g$. Furthermore, if the object is dropped, that means that the initial velocity is zero.

> **Phun Phacts**
>
> Whenever you read a physics problem and find that an object is dropped or is initially at rest then you are given that $v_0 = 0$.

Let the magnitude of displacement be represented by y since the motion is either up or down. The distance is calculated as $y = \bar{v}t$, the average speed is $\bar{v} = \dfrac{v_f + v_0}{2}$, and the acceleration is $-g = \dfrac{v_f - v_0}{t}$.

Weight: The Force Due to Gravity

We calculated the acceleration of a falling object when the force of air friction is negligible. When we made that calculation we divided both sides of the equation by m, as reviewed here: $F_{net} = F_g$, $m\,a_{net} = m\,a_g$, and $a_{net} = a_g$.

This argument and result are correct but you should be aware of the fact that there are two types of mass.

There is gravitational mass that is determined by using a device such as a *triple beam balance*. There is also inertial mass that is determined by a different type of balance called an inertial balance, which measures the amount of opposition to motion an object has due to its inertial mass. It is important to note that inertial mass and gravitational mass are directly proportional. Furthermore, if you choose to measure both with the same unit of mass they are equal.

Newton's second law shows us that mass and weight are proportional but not equivalent. Using that law we can state the magnitude of that relationship symbolically as $W = m_g g$. Again using Newton's second law you find that $W = m_g g$, which means that in the MKS system W is expressed in $\dfrac{\text{kg} - m}{s^2} = N$.

Newton's second law enables us to calculate the force of gravity.

$\vec{F}_g = m_g \, \vec{g}$ By Newton's second law.

$\vec{g} = \dfrac{\vec{F}_g}{m_g}$ Dividing both sides of the previous equation by m.

Remember that the direction of that vector is radially inward toward the center of the circle as indicated in Figure 5.2 for a couple of altitudes above the surface of the earth.

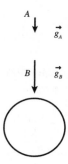

Figure 5.2

The gravitational field vectors represent the gravitational field for two different altitudes.

Imagine concentric spherical surfaces with the vectors $\vec{g}_A$ and $\vec{g}_B$ at every point on the surface and you will get an idea about the gravitational field of the earth for two different altitudes. The units of the gravitational field in each system are $\dfrac{N}{kg}$, $\dfrac{dynes}{g}$, or $\dfrac{lb}{slug}$.

Projectile Motion

The force of gravity affects the motion of every object in motion above the surface of the earth. We may treat as insignificant the force due to air friction by choosing to work with a dense object. Recall the motion of an object that is dropped, $v_{0v} = 0$, described as uniformly accelerated motion with acceleration $-g$.

Motion Horizontally

Suppose we launch an object horizontally with an initial horizontal velocity. In order to distinguish this velocity from any other you may encounter in this discussion, label its magnitude. The motion of this object horizontally is uniform with constant velocity magnitude v_{0H}.

The magnitude of the horizontal displacement is x as has been done before. The magnitude of the horizontal displacement is calculated as $x = v_{0H}t$ where t is the time the object is in motion horizontally. In Figure 5.3, the path of the object is plotted with the two vectors representing its horizontal velocity as the object moves along the path.

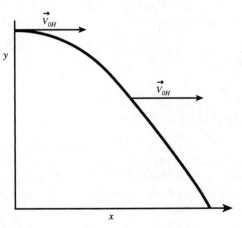

Figure 5.3

The path of an object near the surface of the earth launched horizontally.

At all other points along the path of the object in Figure 5.1, it is the horizontal component of the instantaneous velocity.

Motion Vertically

Let the initial velocity of launch be represented by v_{0v}. The net force is the force of gravity, so the acceleration is $-g$. When the object rises and then stops, that gives us a very important piece of information that we can label $v_{fV} = 0$ and understand it is for the trip upward. Represent the height above the surface of the earth as y. That means $y = 0$ initially or symbolically when $t = 0$. We have enough information to describe the motion:

$y = \overline{v}t$ A good place to begin for uniformly accelerated motion.

$y = v_{0V}t - \dfrac{1}{2}gt^2$ Simplifying, then multiplying by t.

Projectile Path

Actually an ideal projectile undergoes both types of motion simultaneously; uniformly accelerated motion vertically because of effects of gravity and uniform motion horizontally because of zero net force. We describe the path of a projectile by solving the problem here.

A projectile is launched into the atmosphere with an initial velocity $\vec{v}_0$. Find the *time of flight*, *maximum height*, and *range* of the projectile. Let T

represent the time of flight, Y the maximum height, and X the range.

All of these quantities are labeled in Figure 5.4. All of the quantities needed to solve the problem are labeled on that same figure and the information given is listed in the upper-left corner of Figure 5.4.

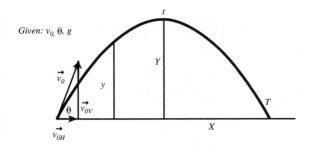

Given: v_0, θ, g

Find: T, Y, X

Figure 5.4

Path of a projectile launched at or near the earth's surface.

The first step in the solution is to determine $\vec{v}_{0H}$ and $\vec{v}_{0V}$ by resolving $\vec{v}_0$ into vertical and horizontal components as shown in Figure 5.4.

Let T represent the time of flight. Let t represent the time to reach maximum height (also the time required to fall from maximum height to the height of launch). Let τ represent any time that the projectile is in flight. Notice that T and t are specific times and τ is a variable that includes those two times. The horizontal distance traveled is $x = v_{0H}\tau$ so half the trip is $x = v_{0H}t$ and for the total trip (range) $X = v_{0H}T$. The vertical distance traveled is $y = v_{0V}\tau - \frac{1}{2}g\tau^2$ so half the trip vertically (the maximum height) is $Y = v_{0V}t - \frac{1}{2}gt^2$.

Since $T = 2t$, we can express the range as $X = 2v_{0H}t$. We found that the maximum height is $Y = v_{0V}t - \frac{1}{2}gt^2$. The problem would be solved if t can be expressed in terms of the information given but $t = \frac{v_{0V}}{g}$.

Now the problem can be solved completely in terms of the information given or implied:

1. $T = 2t$, so $T = \dfrac{2v_{0V}}{g}$.

2. $X = 2v_{0H}t$, and $X = \dfrac{2v_{0H}v_{0V}}{g}$.

3. $Y = v_{0V}t - \dfrac{1}{2}gt^2$, therefore $Y = \dfrac{v_{0V}^2}{2g}$.

The path in Figure 5.4 has the shape of the graph of a parabola.

Deflecting Force

The way gravity acts on the projectile can be explained by resolving the force of gravity vector into two components. One of the components is perpendicular to the instantaneous velocity vector and tends to deflect the moving object out of a straight line path whose direction is that of the instantaneous velocity. The other component is parallel to the instantaneous velocity vector and causes the object to speed up or slow down in the direction of the instantaneous velocity vector.

Consider another deflecting force. Recall that only a centripetal acceleration is associated with an object moving in a circular path at a constant speed. The magnitude of that acceleration is $a_c = \dfrac{v^2}{R}$. It can also be expressed as a vector in the form $\vec{a_c} = -\dfrac{4\pi^2 \vec{R}}{T^2}$ where R is the magnitude of the radius vector and T is the period of the motion. If the object has mass m, Newton's second law enables us to calculate the force that causes that motion. The force is the only force causing the object to move in the circular path at a constant speed and is called centripetal force. The magnitude of centripetal force is $f = ma = \dfrac{mv^2}{R} = \dfrac{m4\pi^2 R}{T^2}$ and has the same direction as the centripetal acceleration, radially inward toward the center of the circular path.

Simple Harmonic Motion

A type of motion that can be related to the uniform motion of an object in a circular path is *simple harmonic motion*.

Imagine that an object of mass m moves with a constant speed around a vertical circular path of radius R as shown in Figure 5.5. As m moves around the circle, the projection of m moves back and forth along the diameter, AB, of the circle. When m starts at A and makes one complete trip around the circle, the projection of m starts at A travels to B and back to A. All of this movement takes place in the period of motion of m, the time to make one complete revolution. The *period* of the motion of the projection of m is exactly the same as the period of the motion of m.

The force on m is the centripetal force labeled $\vec{F}_R$ in Figure 5.5. We think of the horizontal component of the centripetal force labeled $\vec{F}_{RH}$ in that diagram as the force acting on the projection of m. That is, this is a model related to uniform circular motion that can be used to explain another type of motion and its causes. Notice that as m moves counterclockwise from its first position in that diagram, the direction of $\vec{F}_{RH}$ is to the left until m gets to a point directly above the center of the circle. At that instant, $\vec{F}_{RH} = 0$ and as m continues counterclockwise the direction of $\vec{F}_{RH}$ is toward the right and keeps that direction until m is at a point directly beneath the center of the circle where $\vec{F}_{RH}$ becomes zero again When m is directly above the center $\vec{F}_{RH} = 0$, it then increases in magnitude until m is at A where $\vec{F}_{RH}$ is maximum. $\vec{F}_{RH}$ decreases in magnitude, as m continues to move counterclockwise, and becomes zero when m is at a position directly below the center of the circle. It increases in magnitude and becomes maximum magnitude when m is at B. The motion of the projection is said to be periodic with period T, that is, $\vec{F}_{RH}$ goes from zero to maximum to zero to maximum then back to zero while m is making one complete revolution.

The complete trip of m in one revolution and the corresponding trip of its projection is called a *cycle*.

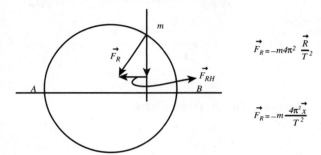

$$\vec{F}_R = -m4\pi^2\,\frac{\vec{R}}{T^2}$$

$$\vec{F}_R = -m\,\frac{4\pi^2 x}{T^2}$$

Figure 5.5

The horizontal component of centripetal force models the restoring force for simple harmonic motion.

Simple harmonic motion is characterized by a restoring force that is directly proportional to the displacement from the equilibrium position and always directed opposite the displacement.

Universal Law of Gravitation

Newton generalized his universal law of gravitation as: any two bodies in the universe attract each other with a force that is directly proportional to the product of their masses and inversely proportional to the square of the distance between their centers of mass. Cavendish determined an accurate value of G experimentally and recently more accurate values of G have be determined as technology for making such measurements have improved. The value of

G is 6.67×10^{-11} and the units were assigned in the same way we assigned units before for the constant in Newton's second law. The universal gravitational constant, G, is $6.67 \times 10^{-11} \dfrac{N - m^2}{\text{kg}^2}$. Newton's universal law of gravitation is state symbolically as $F = \dfrac{Gm_1 m_2}{R^2}$.

The Least You Need to Know

- The centripetal force on an object moving in a circular path at a constant speed is calculated by $\vec{F}_C = \dfrac{-m4\pi^2 \vec{R}}{T^2}$.
- The analysis of the motion of a projectile can determine the time of flight, maximum height, and range of the projectile.
- An object is in free fall if the force of gravity is the only force acting on it.

Chapter 6

Work and Power

In This Chapter

- ◆ Calculating work
- ◆ Applications of work
- ◆ Calculating power

In this chapter, we consider the relationships among three scalar quantities namely kinetic energy, potential energy, and work. Although all are closely related to vector quantities, they are not vector quantities. Momentum and impulse are discussed also and they are both vector quantities.

Work

Two things are required in order to accomplish work in a scientific sense. A force must be applied to an object, and the object must be displaced in the direction of some component of that force.

Work is a scalar quantity. Since work is a scalar quantity, it has no direction associated with it. However, in order to understand what we mean by

work in a scientific sense, we must deal with the
vector quantity of force as well as displacement. A
few examples will help to clear this up.

How Much Work?

In order to calculate the amount of work, we need
a precise definition of work. Work is the result of a
force applied to a body causing the body to undergo
a displacement. The amount of work is calculated
by multiplying the magnitude of the applied force
in the direction of the displacement of an object
by the magnitude of the displacement. It can happen that the force is not applied in the direction of
the displacement. We must avoid the pitfall here of
assuming that work is just force times displacement.
If a force is applied to an object, and the object is
displaced in a direction different from the direction
of the force, then we must multiply the magnitude
of the component of the applied force that acts in
the direction of the displacement by the magnitude
of that displacement. This is shown in Figure 6.1
and Figure 6.2. The work done to move the object
along a horizontal surface in Figure 6.1 may be calculated simply as $W_k = F_A \Delta x$. This is a case where
the force and displacement have the same direction
so the calculation is straightforward.

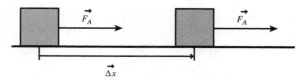

Figure 6.1

The object displaced by a horizontal force is a simple case for the calculation of work.

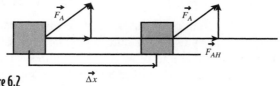

Figure 6.2

The object displaced by a force at an angle with the horizontal is more complicated case for the calculation of work.

The case diagrammed in Figure 6.2 is a little more complicated because as you can see the object does not move in the direction of the applied force $\vec{F}_A$. That means that we must resolve the force into components perpendicular and parallel to the displacement and use the component parallel to the displacement to calculate the amount of work done. The horizontal component of the applied force in Figure 6.2 is labeled $\vec{F}_{AH}$. The work done in that case is $W_k = F_{AH}\Delta x$. All the quantities used in the expression for work are scalars.

Using the expression $W_k = F_{AH}\Delta x$ we will determine the units of work in each system. The MKS system provides the N and m for units of force and displacement respectively so the units for work are $W_k = (N)(m)$ or N–m. Since work is a derived quantity it should be no surprise that the unit of work is newton times meters or N–m. The unit of 1 N–m is defined to be 1 joule so a new defining equation is 1 joule = 1 N–m. Using Newton's second law we know that $1\, nt = \dfrac{1\, kg - m}{s^2}$ so we can express 1 joule in fundamental units as follows: 1 joule = 1 N–m = $1 \dfrac{kg - m^2}{s^2}$. The unit of work in the CGS system turns out to be $W_k = (dynes)(cm)$ or dyne-cm, by using the same algebraic expression as a pattern. The unit of work in the CGS system is the erg. The relationships among all of these units then are expressed as 1 erg = 1 dyne-cm = $1 \dfrac{g - cm^2}{s^2}$. Finally, the unit of work in the FPS system, weird as always, is the ft-lb. There is no one unit of work in the FPS other than 1 ft-lb. The units of work expressed in fundamental units are $1 ft - lb = \dfrac{1 slug - ft^2}{s^2}$.

Example 1: Suppose that a force of 54 N is required to roll a 981 N safe 11 m across the floor. How much work is done? The answer to this problem requires some thought. The way the problem is worded it appears that the force is applied in the direction the safe is moving and that 54 N is just enough force to move the safe at a constant speed. Furthermore, the safe stops when the force is removed. That means that the work done is calculated as follows:

$W_k = F_a \, \Delta x = 54 \ N \times 11 \ m = 590 \ N\text{–}m$ or 590 joules. Unless stated otherwise, it is understood that the object is moved at a constant speed.

Example 2: A force of 68 lb must be applied to the box in Figure 6.2 to move it across the surface in the diagram. The angle at which the force is applied is 42° with the horizontal and the distance the object moves is 12 ft. How much work is done? The same analysis of the situation as in Example 1 is used but we must resolve the force into mutually perpendicular components to solve the problem. The best way to do this is by drawing the force to scale and measuring the angle with a protractor. Construct the projection of the force on a horizontal line to find the horizontal component of the applied force. I urge you to do that and use my results for a check. The horizontal component of the applied force is 51 lb. The solution is …

$$W_k = F_{AH} \Delta x = 51 \ \text{lb} \times 12 \ \text{ft} = 610 \ \text{ft-lb}.$$

Any time something is moved there will be an opposing force, the force of friction. In many cases, it is negligible and you do not have to account for it. In some problems, the force of friction must be calculated. Usually the force applied to get the object to move at a constant speed is equal in magnitude to the force of friction. There is a force of moving friction and a force of starting friction. The moving friction is the force we should be concerned with even though we are well aware that it may be a little less than the force applied initially. The force of friction depends on a lot of things such as the types of surfaces in contact, whether the surfaces are rough or smooth, whether the surfaces are bare and dry or lubricated in some way, and so on. Again we should be aware of how those things can change the force of friction but we can concentrate on the bare and dry surfaces that cause an opposing force.

Sometimes a reference to a table will be given to find the coefficient of friction in when we need it. Once we know the coefficient of friction for two surfaces, the force of friction is calculated as $\Im = \mu N$, and in words it means the force of friction is equal to the product of the coefficient of friction and the normal force. The normal force means the force perpendicular to the surfaces.

Example 3: How much work is done to roll a metal safe, mass 121 kg, a distance of 20.5 m across a level floor? The coefficient of friction is 0.049. We want to move the safe at a constant speed and that means the force applied parallel to the surface must be equal in magnitude to the force of friction. In this case, the normal force is the weight of the safe and we know how to calculate that.

$W_k = F_A \Delta x = \Im \Delta x = \mu N \Delta x = \mu W \Delta x = \mu mg \Delta x.$
The algebraic solution states that the work done is equal to the coefficient of friction times the weight of the safe times the distance moved.
$W_k = (0.049)(121 \text{ kg})(9.8 \text{ m/s}^2)(20.5 \text{ m}) = 1,200$ joules.

Watt Is Power?

Does the power company sell power? Is the more powerful car the one that can lay down the most rubber when it leaves the parking lot? These ideas are discussed within the scientific notion of power. Just as work has a different scientific meaning than the general notion of work, so does power. The first thing we should note is that power is a scalar quantity. There is no direction associated with it.

Power is the work done per unit of time. Some textbooks define power as the rate of doing work. Power depends upon force applied in the direction of displacement, the displacement and time. The definition of power in symbols is $P = \dfrac{W_k}{t}$. As we always do, consider the units of power in all three systems of measurement. In the CGS system, the units of power from the definition are $P = \dfrac{\text{erg}}{\text{s}}$. There is no nice neat unit for power in that system except erg/s. The MKS system unit of power is a little more meaningful $P = \dfrac{\text{joule}}{\text{s}} = \text{watt}$. The units joule/s are always units of power in the MKS system and $1\dfrac{\text{joule}}{\text{s}} = 1$ watt. You are probably more familiar with the kilowatt or 1,000 watts. As usual, the FPS system is weird. The unit of power by the definition is $P = \dfrac{\text{ft}-\text{lb}}{\text{s}}$ in the FPS system. You are probably more familiar with another unit of power in that system, the horsepower or hp. The horsepower is a fairly large unit because

$1 \text{ hp} = \dfrac{550 \text{ ft}-\text{lb}}{\text{s}}$. So if you have 300 horses under the hood of your car, you have a car that will do 165,000 ft-lb of work every second!

Notice the 165,000 ft-lb of work every second. That is what the power company sells in the form of electricity. A kilowatt is a large unit of power and the thing we pay the power company for is work not power. We buy kilowatt-hours that are units of work and not power. Suppose you calculate the work required for 100 watt bulb to provide light in your room. You can say that every second the amount of work is 100 watt-s of work. In an hour, the bulb requires (100 watts) (3600 s) = 3.6×10^5 watt-s = 1×10^2 watt-hr = 1×10^{-1} kilowatt-hr or 0.1 kw-hr.

We can now compare the power of two identical cars. We have decided already that the car that lays down the most rubber when starting off is not necessarily the most powerful. If two identical cars climb the same hill and are timed, the car that gets to the top in the shortest time is the more powerful. They both do the same amount of work on the inclined plane but by the definition of power, the smaller the time the greater the power.

The Least You Need to Know

+ Work is not necessarily force times distance.

+ The units of work are ergs, joules, and ft-lbs.

+ Power is work per unit of time with units ergs/s, watt, and hp.

7

Mechanical Energy

In This Chapter

- Kinetic energy
- Potential energy
- Conservation of momentum

In this chapter we talk about energy and its effects.

Kinetic Energy

A hurricane hitting the coast crashing giant waves into barriers is an example of what scientists call *kinetic energy*. Kinetic energy is the energy an object has due to its motion.

Work and Kinetic Energy

Suppose you observe a bowling ball rolling down an inclined plane. The symbol v' is used to represent final velocity and v is used for initial velocity in the description of motion. The object is undergoing uniformly accelerated motion down the plane.

$\dfrac{v + v'}{2} = \bar{v}$, $\Delta x = \bar{v}t$, $a = \dfrac{v' - v}{t}$ Information used to describe uniformly accelerated motion.

$$\Delta x = \bar{v}t$$
$$\Delta x = \frac{v'^2 - v^2}{2a}$$

$a\Delta x = \dfrac{v'^2 - v^2}{2}$ is obtained by multiplying both sides of the previous statement by a.

The bowling ball has mass, so multiply both sides of the last equation by m.

$$ma\Delta x = \frac{mv'^2 - mv^2}{2}$$

Notice that the left member of the last equation is work, that is, $F\Delta x = \dfrac{mv'^2}{2} - \dfrac{mv^2}{2}$. The quantity $\dfrac{1}{2}mv^2$ is what is called the kinetic energy. That means $W_k = \Delta KE$, where $KE = \dfrac{1}{2}mv^2$.

Example: A crate, mass 2.00 kg, slides from rest down a frictionless inclined plane. The inclined plane is 4.00 m long and makes an angle of 15° with the horizontal.

  a. How large is the force that accelerates the crate down the plane?

  b. How fast is the crate traveling when it reaches the bottom of the plane?

  c. How much time does it take the crate to slide down the plane?

Refer to Figure 7.1 as you consider the solution.

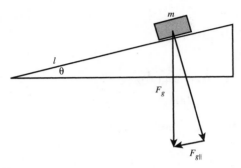

Figure 7.1

The inclined plane for Example 1 includes important labels.

$m = 2.00$ kg $v = 0$

$l = 4.00$ m $\Im = 0$

$\theta = 15.0°$

a. $F_{g\parallel} = ?$

b. $v' = ?$

c. $t = ?$

a) $F_g = mg = 2.00kg \times 9.80\dfrac{m}{s^2} = 19.6N$. Use this value of the weight to construct a vector to represent 5.07 N. When the force triangle is constructed as in Figure 7.1, the geometry of the triangle dictates the length of $F_{g\parallel}$ so that when it is measured using the same scale as used to construct F_g, it will be found that $F_{g\parallel} = 5.07N$.

b) $W_k = \Delta KE$

$F_{g\parallel}\Delta x = \dfrac{1}{2}mv'^2$

$v' = \sqrt{\dfrac{2F_{g\parallel}\Delta x}{m}}$

$v' = \sqrt{\dfrac{2(5.07N)(4.00m)}{2.00kg}} = \sqrt{20.3\dfrac{m^2}{s^2}} = 4.50\,m/s$

Once the final speed is known, t can be determined.

$t = \dfrac{mv'}{F_{g\parallel}}$

$t = \dfrac{(2.00kg)(4.50)\,m/s}{5.07N} = 1.78s$

The Kinetic Energy of a Projectile

Refer to Figure 5.4, where we studied the path of a projectile (see Chapter 5). Suppose that $v_0 = 30.0\,m/s$ at an angle of $40.0°$ with the horizontal and the mass of a potato is 114 g. Calculate the kinetic energy of the potato initially, at maximum height, and just before it touches the ground.

The horizontal component is 23.0 m/s and the vertical component is 19.3 m/s.

The kinetic energy at maximum height is then

$KE = \dfrac{1}{2}(0.114\,kg)(23.0\,m/s)^2 = 30.2\,joules$

The kinetic energy initially is

$KE = \dfrac{1}{2}(0.114\,kg)(30.0\,m/s)^2 = 51.3\,joules$. That

means there was a loss of kinetic energy.

Just before the projectile touches the ground, the instantaneous velocity is −30.0 m/s. The kinetic energy at that time is

$KE = \frac{1}{2}(0.114\,\text{kg})(-30.0\,m\,/\,s)^2 = 51.3$ joules. The

kinetic energy has returned!

Potential Energy

Potential energy is described as the energy an object has as a result of its position in a force field.

In this section, we will choose ground level, or the level with zero potential energy, as our reference point or level.

Work and Potential Energy

The diagram in Figure 7.2 helps to establish that relationship among the quantities work, kinetic energy, and potential energy.

Suppose the object of mass m initially at the foot of the cliff is lifted straight up to the top of the cliff by a force equal in magnitude to the force of gravity on the object. If there is no friction, the stone is raised at a constant speed in order to place it on the cliff. The work done to raise the mass is as follows: $W_k = mgh$ by Newton's second law.

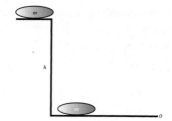

Figure 7.2

The stone and cliff are used in relating potential energy to kinetic energy and work.

Physical Harm

The expression *mgh* is the amount of work done on an object by an applied force upward equal in magnitude to the force of gravity on the object as it is lifted to a height *h* above an arbitrary zero position in the Earth's gravitational force field.

When the object rests on top of the cliff, it has zero kinetic energy. If it is shoved over the side, the object will gain kinetic energy, and just before it touches the ground the potential energy will be zero. At that point the kinetic energy is at a maximum. Of course when it hits the ground both *KE* and *PE* are zero.

It is reasonable to believe that, under ideal conditions, the sum *PE* + *KE* = constant. In this case, the constant is *mgh*. We calculate that sum at one place other than the top and bottom of the cliff to see if that is true.

Suppose that we freeze the object momentarily in space and time the instant it has fallen $\frac{h}{2}$ units, which means that it is $\frac{h}{2}$ units above the ground: $KE = \frac{1}{2}m(gh)$, $KE = \frac{mgh}{2}$, and $PE = mg\left(\frac{h}{2}\right)$ because the object is $\frac{h}{2}$ units above ground.

A High-Flying Potato

Use an initial velocity of 30.0 m/s at an angle of 40.0° again for launching a projectile because some of the arithmetic has been done already. We found $v_{0H} = 23.0 \ m/s \ and \ v_{0V} = 19.3 \ m/s$. We also found that

$$KE = \frac{1}{2}(0.114 \ \text{kg})(30.0 \ m/s)^2 = 51.3 \ \text{joules for the}$$

kinetic energy of the tater when it was launched. In addition, we found that the kinetic energy had become less by the time the potato reached the top of its arc.

Check to see if the sum of the kinetic energy and potential energy at the maximum height is equal to the kinetic energy at launch.

$$KE = \frac{1}{2}(0.114 \ kg)(23.0 \ m/s)^2 = 30.2 \ joules$$

$$PE = (0.114 \ kg)(9.80 \ m/s^2) \ \text{(maximum height)}$$

The maximum height is $Y = \frac{v_{0V}^2}{2g}$.

$KE + PE = 30.2 \ \text{joules} + 21.2 \ \text{joules} = 51.4 \ \text{joules}$

Impulse and Momentum

Suppose we use a baseball bat and toss a baseball into the air and then a 4 kg shot. Each ball is probably pushed into the air with the same force and acted on for about the same amount of time. The baseball had a much larger change in velocity than the shot because of their different masses. The idea that is involved here is called *impulse*. The bat gave both balls the same impulse.

In symbols, $\vec{I} = \vec{F}\,\Delta t$, where the direction of the impulse is the same direction as the force applied.

A change in *momentum* usually implies a change in velocity, but keep in mind that the mass can also change as in the case of a rocket launched from the earth.

When an object of constant mass receives an impulse, it undergoes a change in velocity. The magnitude of the change in velocity depends upon the mass of the object. The physical property that is involved here is called momentum.

Momentum is a vector quantity, and its magnitude depends on the mass and change in velocity of the object being described. Its direction is in the direction of the velocity vector. Momentum is important enough to have a special symbol, $\vec{P}$, assigned to it. Momentum is calculated as $\vec{P} = m\,\vec{v}$.

When an object is given an impulse, it undergoes a change in momentum. The mathematical statement of this relationship is $\vec{F}\,\Delta t = \Delta\left(m\,\vec{v} \right)$.

Physical Harm

Any time an object experiences an
impulse, the object gives an impulse to
the source at the same time. That is, the
bat gives the ball an impulse and the
ball gives the bat an equal and opposite
impulse.

The Relationship Between Impulse and Momentum

Our first look at the relationship between impulse
and momentum will involve expressions modeled
after $\vec{F}\,\Delta t = m\Delta\,\vec{v} = m\left(\vec{v'} - \vec{v}\right) = m\,\vec{v'} - m\,\vec{v}$. They
are involved in motion that is not along a straight
line. Now we can see why the units of impulse are
the same as the units of momentum even though
the units for impulse are N–s and for momentum
$\frac{\text{kg}-m}{s}$ in the MKS system.

Conservation of Momentum

Suppose that we have two steel balls on a table,
and roll one of the balls, m_1, toward the other ball,
m_2, that is at rest initially. Eventually they collide.
Most people have done that experiment many times
before but probably either with marbles or billiard
balls. You know the balls fly off in different directions if there is a glancing blow. Another way of
stating the same thing is the sum of the momenta

after the interaction is equivalent to the sum of the momenta before the interaction.

This is a statement of the law of the conservation of momentum. This law applies even if the interacting bodies stick together after the interaction; momentum is still conserved. You have probably seen cases like this when freight cars are being joined to form a train. One car is usually pushed toward another; the two interact and lock together, then move off together at a slower speed than the first car had initially.

The collision of steel balls is a highly elastic collision and the collision of the freight cars is an inelastic collision. If an object interacts with another object in such a way that the force of interaction depends only on the separation of the objects, the objects are elastic. A highly elastic object may be slightly deformed momentarily but is immediately restored to its original shape; and an inelastic object may be permanently deformed and never return to its original shape.

This business of elasticity is important because kinetic energy is conserved in elastic interactions. It is not conserved in inelastic interactions. Momentum is conserved in all interactions, but not kinetic energy.

The Least You Need to Know

- The change in kinetic energy is equal to the work done.
- Potential energy is the energy an object has due to its position in a force field.
- Impulse causes a change in momentum.

States of Matter

In This Chapter

- ◆ Understanding solids
- ◆ Particles in liquids
- ◆ Studying gases
- ◆ Considering Boyle's law

In this chapter, we emphasize the properties of the different states of matter that lead to applications in physics.

Solids

In solids, the particles are thought to vibrate randomly about a point of equilibrium much like the simple pendulum or the vibrations of a spring. The amplitude of the movement of the particles is quite small as you might expect if you think of tiny springs attached to each particle. The particles are about 0.40 nanometers apart and vibrate with periods of about 10^{-13} s.

The *cohesive forces* between the particles are the causes of the vibrating motion. However, since the distances moved by the particles are small, that force is strong enough to cause the particles to stay together and hold a definite shape for the solid. There are also *adhesive forces* involved when solids of different kinds come in contact. For example, when you accidentally scrub your shoe against the wall some of the finish on your shoe adheres to the wall.

If you lay a bar of gold on top of a bar of lead, you find after a long period of time some gold particles in the lead and some lead particles in the gold. That process is called *diffusion*. Diffusion in solids occurs slowly. Since the particles cause the solid to have a definite shape, the particles also have a definite volume. That means that the solid has mass and inertia. If it is within a gravitational field it will also have a defined weight. Because a solid has definite weight and mass in a definite volume, it has a definite *density*.

We discuss density in a little more detail because density is a general property of all types of matter. Density is a derived quantity and it is a scalar quantity. These symbols are used to represent weight density and mass density: ρ_w for weight density and ρ_m for mass density. The definition of each then is: $\rho_w = \dfrac{w}{v}$ and $\rho_m = \dfrac{m}{v}$. The relationship between the two follows from Newton's second law:

$$\rho_w = \frac{w}{v} = \frac{mg}{v} = g\rho_m.$$

A quantity that is closely related to density is *specific gravity*. In symbols, specific gravity is expressed as $spgr = \dfrac{\rho_{ws}}{\rho_{ww}} = \dfrac{\rho_{ms}}{\rho_{mw}}$. These ideas will serve you very well in science and calculus as well as in the development of ideas in this part of the book.

You should note that weight density, mass density, and specific gravity are all scalar quantities and are derived quantities. Specific gravity has no units of measurement since it is a ratio of two quantities having the same units of measurement. The units of weight density are $\dfrac{N}{m^3}$ in the MKS system, $\dfrac{dynes}{cm^3}$ in the CGS system, and $\dfrac{lb}{ft^3}$ in the FPS system. Remember that some solids like steel springs display the property of elasticity. Elasticity in a steel spring is the ability to stretch or distort within reason and then return to the original length. Elasticity depends on the cohesive forces of the particles of the solid.

Hooke's law is the basic idea of elasticity used in the discussion of the stretching of a spring. The force required to stretch the spring is directly proportional to the elongation of the spring as long as the elastic limit of the spring is not exceeded. If a spring is suspended vertically, it can be calibrated to measure weight. Suppose that you have a spring supported vertically and hang 4.9 N of weight on the spring and observe an elongation (or stretch from the original length) of 0.14 m.

Using Hooke's law, what can be expected for the elongation to be for 9.8 N? We can make this statement using Hooke's law: $F = ky$, $\dfrac{4.9 \text{ N}}{0.14 \ m} = \dfrac{9.8 \text{ N}}{y}$, where y is the unknown elongation. This statement is identified a proportion. Now solve the proportion for y. $y = \dfrac{9.8 \text{ N}}{4.9 \text{ N}}(0.14 \ m) = 0.28 \ m$. What is the elongation for 14.7 N? What is the spring constant in the equation $F = ky$? The spring constant is $35 \dfrac{\text{N}}{m}$ and the elongation is 0.42 m.

Liquids

Like a solid, the particles of liquids vibrate rapidly, but in a more random way. Unlike the vibration of the particles of a solid, which vibrate about a point of equilibrium as if the solid particles are tied to that point by little elastic springs, the particles of the liquid vibrate in all directions haphazardly. The particles of a liquid are about as close to each other as those of a solid. However, they have more freedom of movement around and over each other, with enough mobility to flow and take on the shape of their container. Liquids diffuse just as solids do, but whereas just a few particles of solids move from one solid to another, practically all of the particles of both liquids will diffuse.

The particles of liquids cause cohesive forces as well as adhesive forces much like particles of a solid, except it is found that the forces are not as strong in liquids. As you observed with the stretch of a spring, a considerable force is required to stretch the spring, and even more force is needed to exceed the elastic limit of the spring or to break it. You can observe the strength of the cohesive forces of a liquid by poking your finger in a thick liquid such as paint or motor oil. When you remove your finger, you experience a noticeable tug of the paint particles on your finger (the attraction of paint particles to paint particles) and the paint continues to adhere to your finger even after wiping (the attraction of paint particles to your finger).

The relative strengths of the cohesive and adhesive force of water can be observed by pouring water into a graduated cylinder. The water particles tend to attract the glass particles more than they attract other water particles and cause a concave-up or crescent-shaped water surface. The water particles creep an observable distance up the sides of the cylinder, causing that part of the water surface to be higher than the middle of the column of water. If a column of mercury in a glass tube is observed, it is found that a convex surface on the top of the column forms because mercury particles attract each other more than they attract glass particles. The middle of the mercury surface is noticeably higher than the sides touching the glass.

The cohesive forces of particles in a liquid cause a liquid to have a free surface that is characterized by *surface tension*. Floating a needle on the surface of water can be observed easily demonstrating the surface tension of water. Place a drinking glass filled with water on a level surface. Use an eye dropper to fill the glass until the surface of the water is actually above the glass rim. When the surface is perfectly still, gently place a needle lengthwise onto the surface of the water in the glass. If the surface is not pricked by either end of the needle, it floats!

When the needle is placed on the inflexible film-like surface, it makes a dimple in the surface, increasing the area. The cohesive forces tend to restore the minimum area, a taut horizontal surface, by exerting an equal but opposite force upward. Small bugs use this phenomenon to walk on lake water when it is still. Surface tension also causes small amounts of water such as dew or rain to form small spheres. The spherical surface is found to be the minimum surface for a given volume. The particles of the liquid attract the particles at the surface, causing the surface to have the minimum surface possible with the appearance of an inflexible film stretched over the droplet.

Liquids share another property with solids; they exert *pressure*. Pressure is not force, it is force per area. It is a scalar quantity with units of measurement like pounds per square inch (psi).

Many people use the words *force* and *pressure* interchangeably. There are some distinct differences in the two concepts. To begin with, force is a vector quantity and pressure is not. Next, look at the definition of pressure in symbols, $p = \dfrac{F}{A}$, and determine the units of measurement of this new quantity.

In the FPS system, pressure is measured in $\dfrac{lb}{in^2}$ sometimes referred to as psi. The units of measurement in the MKS system are $\dfrac{N}{m^2}$.

An example of the type of problem you can encounter involving pressure might be helpful here, especially if it emphasizes the difference in force and pressure.

Example: Suppose that a 110-lb woman walks into the room wearing, among other things, high-heeled shoes. The heels have an area of about $\dfrac{1}{16}$ in^2 at the base. Her full weight is first on one heel and then the other as she walks across the floor. How much pressure does she exert when her heels touch the floor?

Solution: $p = \dfrac{F}{A} = \dfrac{110 \, lb}{\dfrac{1}{16} \, in^2} = 1{,}760 \, lb/\, in^2!$ The correct answer is 1,800 lb/in^2. That means that in this case a force of 110 lb causes a pressure of 1,800 lb/in^2. There is also a difference in units of measurement as well as the fact that force is a vector quantity and pressure is a scalar quantity.

Gases

The particles of a gas are so far apart that they do not exert forces on each other until they bump into each other. Obviously, the particles are very far apart compared those of solids and liquids. In fact, particles of matter in the gaseous state occupy about 10^3 times the volume that they occupy in the liquid state. That does not mean that the particles are bigger now that they are gaseous, just that they take up more space.

At room temperature, gas particles travel at about 500 m/s and travel about 100 nanometers before bumping into another particle or the sides of the container such as a sealed balloon or glass beaker. The sides of the container are experiencing between 4 and 10 billion collisions each second. Gases have density, exert pressure, and diffuse. Oh, do they diffuse! If someone walks into the room wearing wild perfume, you smell it almost immediately. The gas particles move randomly in all directions. They mix with the other gas molecules in a container, the room. These particles move so fast and are so far apart that we can think of them as individual particles. It should be obvious by now that a gas does not have a definite volume or a definite shape. The particles move so fast in a random fashion that they completely fill their container.

Physical Harm

Even people who should know better confuse the words force and pressure. Remind yourself, with notes if necessary, that force is a vector quantity and pressure is a scalar quantity. Be able to identify the units of both in all systems of measurement.

The density of a gas is the same quantity as the density of a solid or liquid. Air has mass and weight. The mass density of air near sea level is found to be 1.29×10^{-3} g/cm^3. We are fortunate to live on a planet that surrounds us with a sea of gases, the atmosphere. The atmosphere that I am referring to is at or near sea level, but the atmosphere of the earth extends up for about 25 miles although most of it is below an altitude of 20 miles above the surface of the earth. Above 25 miles the atmosphere gets pretty thin. The pressure of the air we breathe near the surface of the earth is often reported as the pressure of a column of air one square inch in area and reaching to the top of the atmosphere.

The pressure at sea level of our atmosphere is 1 atmosphere, or 1 atm. That is the same as 14.7 lb/in^2 = 1.013×10^6 dynes/cm^2. There are other measures of pressure, and some of them are listed here for your information: $1 Pa = 1 \text{ Pascal} = 1\dfrac{N}{m^2}$, $1\, bar = 1.000 \times 10^5\, N\,/\,m^2 = 10^5\, Pa$.

Physical Harm _____

Unless stated otherwise in a problem, the pressure near the earth's surface means the pressure at sea level at standard temperature 0°C.

The unit of measure of pressure used depends on the application. The physicist will use the units reviewed earlier. The chemist will use the same units as the physicist along with millimeters of mercury, or Torr. The weatherman uses millimeters of mercury, inches of mercury, and bars or millibars when he discusses pressure in his weather reports.

Boyle's Law

Two reasons for considering Boyle's law are (1) to emphasize the limitations of a mathematical model borrowed from mathematics by the scientist, and (2) to explore the relationship of two properties that we have discussed for all states of matter: volume and pressure. Recall that the particles in solids and liquids attract each other at certain distances, and if the distances are less the forces become repulsive. That means that solids and liquids cannot be compressed easily. The particles in a gas are so far apart that gases have no fixed shape or volume. The particles of a gas can be made to occupy a smaller volume with relative ease.

The particles of solids differ from liquids and gases in that the particles of solids vibrate randomly about a point of equilibrium. The particles of solids do not have the freedom of motion that the particles of liquids and gases do. Because of the freedom of motion of the particles of liquids and gases, liquids and gases are referred to collectively as fluids. A known mass of gas in a closed container and constant temperature will behave according to the mathematical model shown by the graph in Figure 8.1. Using the graphical analysis of data outlined in an earlier chapter you can see that Figure 8.1 suggests an inverse variation between the volume of a gas and the pressure that it exerts if the temperature and amount of gas particles remain constant.

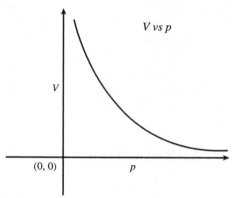

Figure 8.1

A graph of volume vs. pressure reveals a mathematical model called Boyle's law.

In fact, as the data is presented in that graph, we can see that apparently $V \propto \dfrac{1}{p}$. We can see just how closely the data fits that mathematical model by plotting V vs. $\dfrac{1}{p}$. A straight-line graph reveals that you have a good model to fit the data, and in addition the slope of the graph is the constant of proportionality. It will be no surprise, after your work with this book, to find that Boyle's law is not the absolute truth. That is, it is not always true. You know that if the pressure is increased enough so that the particles start affecting each other by attraction the gas starts behaving more like a liquid and less like a gas.

When you see a mathematical statement of a scientific law or principle, you know that the statement does not mean that the law or principle is true now, always has been true, and always will be true. The scientist identifies models that best fit the observations of his or her experimentations with the constraints of his or her laboratory procedures. It is important to keep the limitations of science in mind.

The Least You Need to Know

- ◆ Solids have a definite shape.
- ◆ Liquids assume the shape of their container.
- ◆ A gas rapidly fills its container completely.
- ◆ Boyle's law states that the volume of an ideal gas varies inversely with the pressure.

Pressure

In This Chapter

- Pressure of liquids
- Pressure of gases
- Buoyant force

Ships at sea obey the same laws of physics to haul cargo from one continent to another. In this chapter, we look at why a barge or ship must sink far enough into the water to support the craft and cargo.

Pressure and Liquids

Pressure is defined to be force per unit area. Pressure is exerted in all directions in a fluid. Suppose you have a confined fluid, oil maybe, in a container with a cylinder at each end both completely filled with the fluid like that found in Figure 9.1.

On top of each cylinder is a piston that is free to move up and down. This arrangement constitutes a simple machine that will multiply force by using pressure on the confined motionless fluid. The machine is based on Pascal's principle which states that any pressure applied to a confined fluid at rest will be transmitted undiminished to every point in the fluid. So if you push down on A_E the pressure transmitted undiminished throughout the fluid will cause a force upward on A_R. This arrangement acts like a simple machine and has applications like the lift that raises your car when the mechanic changes the oil or switches the tires.

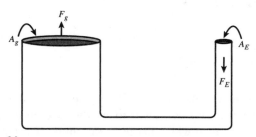

Figure 9.1

The hydraulic press operates as a simple machine.

Physical Harm

Your mechanic probably applies the effort force by releasing compressed air into a chamber that causes the effort piston to move. The details discussed here are in play at the service station but all you see the giant cylinder move upward raising your car on the service ramp.

Liquid Pressure and Depth

Refer to Figure 9.2. That diagram is used to develop a way of calculating the pressure at any depth of a liquid.

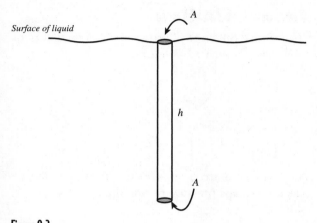

Figure 9.2

Calculating the pressure on the base of a cylinder of liquid.

The top of the cylinder is at the surface of the liquid. The bottom of the cylinder is at a depth h in the liquid. The cylinder is just a cylinder of the liquid itself. The pressure on the bottom of the cylinder is the weight of the liquid in the cylinder divided by the area of the base of the cylinder. The cylinder is uniform, with the area of the base the same as the area of the top of the cylinder. The pressure on the bottom of the cylinder is calculated in the following way:

$$p = \frac{\rho_L V_L}{A}$$

$$p = \frac{\rho_L A h}{A}$$

$$p = \rho_L h.$$

Pressure and Total Force

The total force on the bottom of a rectangular pool of water is calculated as follows:

$$p = \frac{F}{A}$$

$$F = pA$$

$$F = \rho_w h A$$

$$F = \rho_w V.$$

Since the pressure varies with the depth, you must use the average pressure to find the total force on a side.

$$p_{avg} = \frac{F_{total}}{A_{side}}$$

$$F_{total} = p_{avg} A_{side}$$

$$F_{total} = \rho_w h_{avg} A_{side}.$$

Pressure and Buoyant Force

Refer to Figure 9.3 and think of the small uniform cylinder as completely submerged in a liquid.

Use water for the liquid. Due to the pressure of the liquid, there is a force, f_2, acting upward on the bottom of the submerged cylinder that is at a depth h_2. For the same reason, there is a force, f_1, with direction downward on the top of the submerged cylinder. Since pressure varies directly as the depth of a liquid, we know that $f_1 > f_2$, so there must be a net upward force due to the liquid.

$$f_2 - f_1 = f_{net}$$
$$\rho_{ww}h_2 - \rho_{ww}h_1 = f_{net}$$
$$\rho_{ww}(h_2 - h_1) = f_{net}$$

$\rho_{ww}h = f_{net}$ Because $h_2 - h_1 = h$, the height of the submerged cylinder.

$$W_{liquid\ displaced} = f_{buoyant}$$

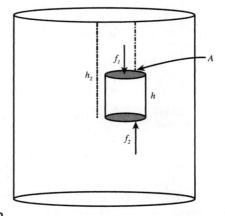

Figure 9.3

The cylinder completely submerged in water has a buoyant force.

Suppose the object is not completely submerged. We know that the object must float in the liquid and is not moving vertically. That means that there is no net force on the floating object. Refer to Figure 9.4 to see a cylindrical object that has length, h, and uniform cross section, A_{cyl}, floating in a liquid with h_a out of the liquid and h_L submerged in the liquid.

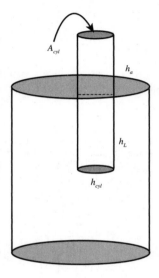

Figure 9.4

The floating cylinder in a liquid that provides a buoyant force.

The buoyant force on the object must have the same magnitude as the weight of the object. That is, if the liquid is water,

$$f_b = W_{cyl}$$

$$\rho_{ww} V_{w\ disp} = \rho_{w\ cyl} V_{cyl}$$

$$\frac{h_L}{h_{cyl}} = \frac{\rho_{w\ cyl}}{\rho_{ww}}$$

Phun Phacts

The LST is made of steel and so are submarines and other ships, and yet they float very well. It does not matter what an object is made of as long as it displaces enough water so that the buoyant force of the water is greater than the weight of the object floating in the water.

A gas exerts pressure but in a different way than a liquid. The particles of the gas collide with the sides of the container many times each second. In each collision, the particles and sides exchange impulses. The pressure of a gas is the result of the billions of impulses of the bouncing particles.

Example: The atmosphere (atm) is a good source of gas, or a mixture of gases, with which to begin. We found earlier that at sea level the pressure of the atmosphere is 14.7 lb/in^2.

Soda is sipped through a straw by removing the air from the straw. The liquid is pushed up to the mouth by the atmospheric pressure on the surface of the liquid outside the straw. That suggests an idea for measuring the pressure of the atmosphere.

We can measure the pressure of the atmosphere by relating it to the pressure of a liquid; mercury. The height of mercury in a tube stands at 760 mm, 76.0 cm of Hg or 29.92 in of Hg when the measurement is made at sea level and when the temperature is 0°C. When you work with gases, the conditions STP (standard temperature and pressure) are 76.0 cm of mercury and 0°C.

The mercurial barometer is diagramed in Figure 14.5, showing that 1 atm of pressure on the reservoir of mercury causes the mercury to rise to a height of 76.0 cm inside the glass tube.

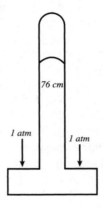

Figure 9.5

The mercurial barometer measures the atmospheric pressure directly.

Mercury is a liquid that is readily available and has the largest density of common liquids. If water is used, we find that the height of water would be about 34 ft for standard pressure, or 13.6 times higher than the column of mercury.

$$p = \rho_{ww}h_w = \left(62.4 \text{ lb/ft}^3\right)\left(34 \text{ ft}\right) =$$
$$\left(2122 \text{ lb/ft}^2\right)\left(\frac{1 \text{ ft}^2}{144 \text{ in}^2}\right) = 14.7 \text{ lb/ in}^2$$

$$p = \rho_{wm}h_m = \rho_{ww}spgr_m h_m =$$
$$\left(980 \text{ dynes/cm}^3\right)\left(13.6\right)\left(76.0 \text{ cm}\right) =$$
$$= 1.013 \times 10^6 \text{ dynes/cm}^2$$

Phun Phacts

Buoyant force is calculated for gases the same way that it is in liquids. Since they are both fluids you can say the buoyant force is the weight of the fluid displaced by the object immersed in the fluid.

$FB - W_{He} = F_L$ In words, the equation states that the buoyant force minus the weight of helium, actually helium and container, is equal to the lifting force (helium in this case).

$$\rho_{wA}V_{He} - \rho_{He}V_{He} = F_L$$
$$(11)\left(1.29\frac{\text{g}}{\text{l}} \times 980\frac{\text{cm}}{\text{s}^2}\right)(1 - 0.138) = F_L$$

$$F_L = 1.09 \times 10^3 \text{ dynes}.$$

That means that one liter of helium will lift 1090 dynes of weight at STP.

The Least You Need to Know

- Standard temperature and pressure is 0°C and 76 cm of mercury.

- The pressure due to a uniform liquid at any depth is pressure equals weight density of the liquid times the depth.

- The buoyant force of a fluid is the weight of the fluid displaced.

Heat

In This Chapter

- Measuring temperature
- Degrees Celsius or Celsius degrees
- Making work easy
- Quantities of thermal energy

Something is hot or cold depending on its thermal energy. The thermal energy of a body is the kinetic energy of the particles of the body. A hot body can transfer heat to a cold body and when it does, its thermal energy is reduced. The thermal energy of the body absorbing the heat is increased.

Temperature is used to identify a body that is hot or cold. In this chapter, I review the use of three different scales for measuring temperature. Ideally thermal energy is conserved in the transfer of heat from an object with a higher temperature to an object with a lower temperature.

Temperature

When a pan of water is boiled, the water gets hot. If some of the hot water is poured into a cup and the pan is set on a hot pad near the cup, the cup of hot water does not have as much *heat* as the several cups of hot water left in the pan.

If the agitation of those particles of matter is increased, their total internal energies are increased so the body containing those particles has more thermal energy. The motion of the particles can be increased by causing a chemical reaction like striking a match, by placing them so that they are exposed to a radiant source like the sun, by bending a piece of wire back and forth several times, or by passing an electrical current through wire like the filament of a light bulb.

Thermometers

Temperature can be measured with the proper instrument but thermal energy is more difficult to determine quantitatively. Actually how much thermal energy is in something is measured only by how much is transferred. Consider the measurement of temperature first and then we will spend some time calculating the quantity, Q, the symbol for the quantity of thermal energy. The thermometer is the instrument for measuring temperature. There are four different scales for measuring temperature. Three of those scales are discussed here. Most people in the world are interested only in the absolute or Kelvin scale and the centigrade or Celsius scale. The Fahrenheit scale is included because that is the scale used widely for the measurement of temperature in the United States. Readings on the Celsius scale are written as 20°C, on the Fahrenheit 20°F, and on the Kelvin scale 20 K. These three temperatures are not the same. You would be comfortable at 20°C, freezing at 20°F, and dead stiff at 20 K.

Effects of Change in Temperature

When heat is added to most materials, they expand. Glass and mercury expand when heated but mercury expands more than glass. The thermometer is calibrated to read temperature as the difference in the expansion of mercury and glass.

Physical Harm

When liquid water releases heat, it becomes denser until it reaches about 4°C, and then gets less dense as it continues to release heat. As water changes to ice, it continues to release thermal energy. The ice expands with such a tremendous force that it can break the blocks of cars that are not properly protected by antifreeze. The expansion is so drastic that ice floats in water, though most of the ice is below the surface. This is not typical of most solids in contact with their liquids. Most materials are denser in their solid state than in their liquid state.

Water is the standard for calibrating the thermometer because it has a definite point at which it freezes, called the *freezing point of water*, and at which it boils, at standard pressure, called the *boiling point of water*. The procedure used to calibrate a thermometer is to place the thermometer in water and let the water freeze. Then etch on the glass the place matching the top of the mercury column. Next place the thermometer in water and let it boil, at standard pressure. Finally, etch on the glass the place matching the top of the mercury column.

Of course, today we use alcohol in many thermometers rather than mercury to avoid the problems that come from broken thermometers and escaping mercury.

Comparing Temperature Scales

The freezing point on the Celsius scale is named 0°C, on the Fahrenheit scale it is designated 32°F, and on the Kelvin scale it is named 273 K. The boiling point on the Celsius scale is labeled 100°C, on the Fahrenheit scale it is labeled 212°F, and on the Kelvin scale it is designated 373 K.

Temperature Change and Change in Temperature

Observe that the thermometer reads 0°C when the bulb of the thermometer is in freezing water and that it reads 100°C when placed in boiling water at standard pressure. Another use is to observe a temperature change from a reading of 0°C to a reading of 100°C, for example. That means that we observed a change in temperature of $\Delta t = 100°C - 0°C = 100C°$. Notice the difference in reporting these two observations; the first observation is two different readings and the second observation is 100 Celsius degrees change in temperature.

Physical Harm

Converting a Celsius reading to a Kelvin reading involves the calculation of a change in temperature from the freezing point to the reading to be converted. The change in temperature is in C°. The C° is equal to the K degree so adding the change in temperature from the freezing point to the reading on the Celsius scale to the freezing point on the Kelvin scale results in the correct Kelvin reading. The same procedure is followed when converting from Kelvin to Celsius readings.

Phun Phacts

One way of thinking about converting readings between Celsius and Fahrenheit scales is to realize that $100C° = 180F°$ and $5C° = 9F°$. That means, to convert $25°C$ to a Fahrenheit reading, realize that the Celsius reading is $25C°$ from the freezing point. That is the same as $\left(25\left(\dfrac{9}{5}\right)\right)F°$ or $45F°$ from the freezing point or a reading of $(45 + 32)°F = 77°F$.

Specific Heat

Temperature is the condition of a body to transfer thermal energy to another body or to absorb thermal energy. The measure of the temperature of a body is needed to find a quantity of thermal energy. The body has other properties that must be taken into account when we determine the quantity of heat of a body.

First, the calorie (cal) is the quantity of heat required to raise the temperature of one gram of water one Celsius degree. This unit of heat is sometimes referred to as the small calorie to emphasize the difference between the calorie and the Calorie, or so-called large calorie. The MKS unit of heat is the kilocalorie (kcal), which is also referred to as the Calorie, the unit of heat used by dieticians and biological scientists to measure the fuel value of foods. As is usually the case for the FPS system, the unit of heat is strange; it is the Btu. The British thermal unit (Btu) is the quantity of heat required to raise the temperature of one pound of water one Fahrenheit degree.

The relationship between work and heat that was developed by James Prescott Joule is summarized with the equation $W_k = JQ$. From that relationship, we find that 1 cal = 4.19 joule and 1 kcal = 4,190 joules. It is helpful to know a relationship between units of thermal energy in the FPS system and the metric system, as well as SI units, 1 Btu = 252 cal = 1,056 joules. The contractors who discuss the needs of your home for air conditioners and furnaces usually talk about the number of Btus involved.

Objects that take longer to increase their temperature also take longer to cool off than others do. This property of a substance is called its heat capacity. The heat capacity of a substance is the heat necessary to raise the temperature of a material one degree. The heat capacity of an object can be expressed as: heat capacity $= \dfrac{Q}{\Delta t}$, with units of a more practical property of a substance that takes into account the type of substance and mass is the *specific heat* of an object. The specific heat depends on the particular type of matter as well as the mass of the object. Specific heat is usually represented as: $c = \dfrac{Q}{m\Delta t}$ with units $\dfrac{\text{kcal}}{\text{kg C}^\circ}, \dfrac{\text{cal}}{\text{g C}^\circ}$, or $\dfrac{\text{Btu}}{\text{lb F}^\circ}$.

We now have one way of calculating a quantity of thermal energy. Suppose we are given a 5.00 g piece of aluminum, $c = 0.214 \dfrac{\text{cal}}{\text{g C}^\circ}$, that is warmed up from 5.00°C to 55.0°C. How much heat is required to cause the change in temperature? The specific heat gives you a start: $C_{AL} = \dfrac{Q}{M_{AL}\Delta t}$, $Q = C_{AL}M_{AL}\left(t_f - t_0\right)$, and $Q = \left(0.214 \dfrac{cal}{g\,C°}\right)(5.00\ g)(55.0°-5.00°C) = 53.5\ cal$.

The problem just completed had the change in temperature as part of the information given. Thermal energy leaves objects of higher temperature and enters objects of lower temperature until their temperatures are the same. A symbolic statement of the law is Q_{gained} by one object = $Q_{released}$ by the other.

The thermal energy will have been given up to the outside air if it is not isolated from the environment. In an isolated system, any thermal energy that is lost by hot objects or substances is gained by cold objects or substances. An application of this law will help you see how this works.

Latent Heat of Vaporization and Fusion

One method of calculating a quantity of thermal energy involves a change of state. Thermal energy can be added to a solid and the temperature of the solid rises until it reaches a certain characteristic temperature. At that temperature, the solid changes to a liquid. The temperature at which this happens is called the *melting point*. The thermal energy continues to be absorbed, but the temperature does not increase. Work is being done. The internal potential energy of the material is increasing just like it would if you pulled on a spring and stretched it. The process of changing from a solid to a liquid is called *fusion*. Crystalline solids, like ice, have a definite melting point, and the *freezing point* is the same temperature. A liquid becomes a solid during a process called *solidification*.

The temperature at which solidification takes place is called the freezing point. The melting point of ice is 0°C and the freezing point of water is 0°C but even though the temperature is the same the processes are very different.

In order to change ice from your freezer to a liquid, in other words to melt ice, heat must be added to the ice until it is at 0°C; then we must continue to add thermal energy to the ice until it changes to water at 0°C. The amount of thermal energy that must added to ice at 0°C to change it to water at the same temperature is called the latent heat of fusion, or just the heat of fusion. Each crystalline solid has a characteristic heat of fusion. The heat of fusion for ice is $80\dfrac{\text{cal}}{\text{g}}$ or $144\dfrac{\text{Btu}}{\text{lb}}$.

In like manner, water must be cooled until it reaches 0°C, and then 80 cal of thermal energy must be removed from each gram of water to change it to ice at 0°C. The specific heat of ice is used to find the quantity of heat necessary to warm it up to the melting point then calculate the quantity of thermal energy necessary to change it to water by using the heat of fusion. Just the reverse of this procedure is followed to change water to ice.

Physical Harm

> The heat of fusion is the quantity of thermal energy required to change a solid to a liquid or a liquid to a solid without changing the temperature of the final substance.

A liquid must absorb heat until it reaches its boiling point. That requires the use of the specific heat of the particular liquid. Then we must continue adding thermal energy to the liquid at the boiling point until the liquid changes to a gas or a vapor at the boiling point. The process is called *boiling*. Changing from a gas to a liquid is just the reverse process and is called *condensation*.

The gas is cooled off until it reaches the boiling point. Calculating the quantity of heat removed from the gas to cool it to the boiling point requires the use of the specific heat of the gas. By continuing to remove thermal energy from the gas or vapor at the normal boiling point (the boiling point at standard pressure), we can cause it to condense, and it becomes a liquid at the boiling point.

The quantity of thermal energy required to change a liquid to a vapor or a gas is called the latent heat of vaporization, or *heat of vaporization*.

The heat of vaporization is the quantity of thermal energy required to change a unit mass or weight of a liquid to a vapor or gas at the normal boiling point. The heat of vaporization for water is 540 cal/g or 970 Btu/lb. That means that once the temperature of water is increased to the normal boiling point of 100°C we must continue adding thermal energy to change the water to steam at the same temperature.

Heat and Work

A source of heat is placed in a Styrofoam (Styrofoam specific heat about the same as water) cup of water for a measured time. The equation that relates the reversible equivalence of work and heat is used to calculate the constant $\mathcal{J} = 4.19$ joules/cal. A summary of calculations made is: $W = \mathcal{J}Q$, where W is the work and Q is the amount of heat developed. $Pt = \mathcal{J}(m_{cup+water})c_w(t_f - t_w)$, t_f is the temperature at the end of the time period that the heat source was in contact with the cup of water and t_w is the initial temperature of the cup and water.

$$\mathcal{J} = \frac{Pt}{\left(m_{cup+water}\right)c_w\left(t_f - t_w\right)}, \text{ where } P \text{ is in watts and}$$

t in seconds.

The Least You Need to Know

- ◆ Temperature specifies the condition of a body to transfer thermal energy.
- ◆ Temperature is measured in K, °C, or °F and a change in temperature is measured in K, C°, or F°.
- ◆ Calculating quantities of thermal energy requires specific heat, heat of fusion, and heat of vaporization.

11

Sound

In This Chapter

- ◆ Sources of sound
- ◆ Wave nature of sound
- ◆ Speed of sound

In this chapter, we consider how sound originates in ways other than a singing bird or a barking dog.

Source of Sound

A familiar question is about a tree falling in the forest and whether there is sound resulting from the falling tree if no one is there to hear it. The physiologist might say that in order to have sound there must be a source, a medium to transmit the sound, and a receiver of sound. He may say since there is no receiver, there is no sound. The physicist might say that sound is a special disturbance of matter to which the ear is sensitive. He may say that sound is there whether it is received by an ear or not. Those same special disturbances may also be beyond the ability of an ear to detect them. Consider now those disturbances in matter that we refer to as sound.

As an approach to the study of sound, refer to a topic presented earlier in this book. Recall the information about the simple pendulum.

> ### Phun Phacts
>
> Any object undergoing simple harmonic motion has associated with it not only a period of motion but also a cycle. One cycle starts at equilibrium, goes to maximum displacement in one direction and back to equilibrium, then to maximum displacement in the other direction and back to equilibrium. Remember that the maximum displacement from equilibrium is called the amplitude.

Once set in motion the pendulum moves with simple harmonic motion. The motion is a vibratory motion that is repeated over and over again. The time for the pendulum to complete one cycle is called the period of the motion. The reciprocal of the period is the frequency of the motion. The period is measured in seconds and the frequency in cycles/s or hertz (Hz). The disturbances in matter that our ears are sensitive to are very much like the to and fro motion of the pendulum.

The strip of metal in Figure 11.1 is secured to a desk so that one end is free to vibrate in a vertical plane. The vibrating strip is actually a physical pendulum.

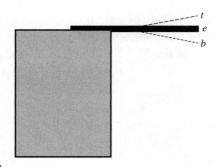

Figure 11.1

The vibrating metal strip compresses air particles on one side and then pulls away from them as it continues to vibrate, leaving a relative void.

Phun Phacts

The frequency of simple harmonic motion is the reciprocal of the period of motion. That is, $f = \frac{1}{T}$ and $T = \frac{1}{f}$ so the period is measured in seconds per cycle or just seconds, and the frequency is measured in s^{-1}, $\frac{1}{s}$, or $\frac{cycle}{s}$.

Vibrating metal is much like the simple pendulum, but it is different in that you can hear the sound associated with the disturbance of the metal strip. Sound can be heard if the strip vibrates with frequencies in the audible range of the ear, about 20/s to 20,000/s.

The vibrating string of a violin or the booming of a kettledrum are other examples of disturbances of matter that we detect as sound.

Types of Waves

The vibrating metal strip transfers mechanical energy to the air particles surrounding it as it moves. When it moves from e to b, the air particles below the strip are pushed together downward, compressing them. That leaves a relative void in the space above the metal strip. The strip has its maximum velocity when it passes through e so it causes maximum compression of air particles below the strip at that time. While moving from e to b, the strip continues to compress the particles but the compression becomes less and less as the speed slows to zero at b. As the metal moves from b to e the particles beneath the strip get farther apart.

Physical Harm

Even though the diagrams of waves are linear, remember that waves can be thought of in two dimensions like water waves or in three dimensions as expanding spherical surfaces.

The particles continue to be separated as the strip moves from *e* to *t*, but they become less and less separated as the velocity slows to zero at *t*. From *t* to *e*, the particles beneath the metal strip are being compressed more and more until maximum compression is reached at *e* when the cycle begins again.

Transverse Waves

The *transverse wave* is probably the most familiar type of wave. At some time you have held one end of a rope in your hand and sent a loop down the rope by shaking your hand perpendicular to the rope. The loop traveled down the rope all by itself without having any particles of the rope making the trip with it.

A wave can be a disturbance that travels through a medium. When the particles vibrate perpendicular to the direction the wave is traveling, the wave is called a transverse wave.

Longitudinal Waves

Sound does not fit the transverse wave model. Sound can be thought of as a *longitudinal wave* because of the vibrations of the particles of the medium.

A longitudinal wave is a disturbance moving through a medium in which the particles of the medium vibrate in paths that are parallel to the direction of travel of the wave.

The pulse you create in the rope by a flick of the wrist and the pulse the metal strip makes in the downward part of a vibration both travel through the media without having matter move along with them.

Properties of Waves

Hold the end of a rope at a position you may call the equilibrium position, and then move your hand upward then back downward to equilibrium to create a pulse that is called a loop. If you repeat that motion but also continue through the equilibrium position downward and then reverse the motion upward to equilibrium, you create the loop followed by a bowl-shaped pulse. If you continue that motion with a fixed frequency for your hand, a train of pulses from your hand to the other end of the rope is created. Similarly, the vibrating strip will send a train of pulses through the air with regions where the particles are compressed followed by regions where the particles are spread apart.

In Figure 11.2, a transverse wave and a longitudinal wave are diagrammed for comparison.

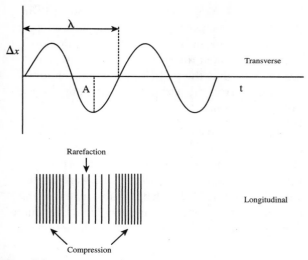

Figure 11.2

The diagram of the transverse and longitudinal waves illustrates their distinguishing characteristics.

First look at the transverse wave that is a graph of the displacement of the tip of the bottom side of the vibrating metal strip versus the time. The amplitude of the displacement, the maximum displacement of the tip, is labeled A in the diagram. The part that looks like a camel's hump is called a *crest* and the bowl-shaped portion is called a *trough*.

The distance from the beginning of a crest to the end of an adjacent trough is the *wavelength*, labeled λ in the diagram, of the wave. When the metal strip is passing through the equilibrium position in Figure 11.1 moving downward a *compression* is being created. As the strip continues down to *b* in that diagram the particles are getting farther apart because the bottom of the metal is moving away from them. When the strip continues from *b* and passes through *e*, the particles are at a maximum distance apart in the part of the longitudinal wave called a *rarefaction*.

We have just traced the creation of part of one complete longitudinal wave that makes up one compression in a train of waves. That part of a wave has a length that is called half a wavelength of a longitudinal wave. It is half the distance that the beginning of the pulse moves away from the source during the time that it takes for the whole cycle to occur.

Phun Phacts

The compressions and rarefactions of a longitudinal wave are regions where particles are closer together and farther apart, respectively. The action of pushing and spreading apart throughout the room from the source of sound is the wave nature of sound. It spreads like ripples on the surface of water (but in three dimensions in a room) that remains circular (or spherical in three dimensions) because the speed of the disturbances is constant.

If there is a good fit, then the model is used to explain the observations of a material or event. If the model not only explains most observations about a phenomenon but also predicts other things that an experimenter may look for relating to his other observations, it is a model that is adopted and used widely. That means that viewing sound as a wave, and not only that but a longitudinal wave, lends itself to better understanding of sound.

Speed of Sound

The wave model provides for several properties of sound. Sound requires a medium for transmission Sound does not travel in air as fast as light travels. Sound does have a finite speed as does light, but over short distances the speed of light is so fast that it seems practically instantaneous.

Phun Phacts

Outer space, like the region above the surface of the moon, is about as close to a vacuum as you can get. Remember that astronauts dropped a feather and a hammer and both hit the surface of the moon at the same time. The hammer and feather experienced no opposition to their motion from air as they fell to the moon's surface. Sound would not be transmitted through that same region because it requires a medium to conduct the sound wave.

Motion and Sound

Most of us have observed the delayed sound of a distant shovel on a cold winter morning. The action is seen and later the result of the action is heard. The flash of lightning in a thunderstorm is seen and the thunder is heard a short time later. Sound does travel with a definite speed that is consistent with the wave model. The speed of sound is about 331.5 m/s at 0°C; that is about 1087 ft/s at 32°F or about 740 mph.

Uniform motion is involved in the description of sound or sound waves. If you are caught out in a thunderstorm and see a flash of lightning, start counting one-thousand-one, and so on, and when you hear the thunder, mentally multiply the number you counted by 1,100 ft/s to get a good idea of how far away the storm is from you. The sound wave is traveling at a constant speed.

Wavelength and Frequency

Suppose we pursue the notion of a sound wave just a bit further. It was suggested that your hand can model the vibrating metal strip by doing several things, one of which was to set your hand in motion at a certain frequency. What do you suppose would happen if you doubled the frequency of your hand on the rope while keeping all other conditions the same? This means you have to maintain the same amplitude and keep the rope at the same tension. The wavelength in the rope will automatically get shorter as the frequency is increased.

What would you expect of your graph if you were to plot the wavelength versus the frequency? This graph is plotted in Figure 11.3.

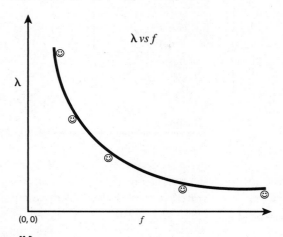

Figure 11.3

The graph of wavelength versus frequency suggests an inverse relationship.

The graph suggests an inverse variation, that is, $\lambda \propto \dfrac{1}{f}$. That is graphed in Figure 11.4, where it can be seen that the graph confirms the notion because the straight line means that $\lambda = \dfrac{k}{f}$ or $f\lambda = k$, where k is the constant of proportionality. The speed is constant in this case.

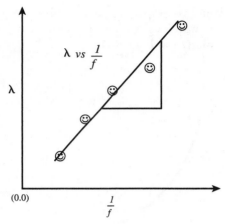

Figure 11.4

The graph of wavelength versus the reciprocal of frequency confirms that $\lambda = \dfrac{k}{f}$.

The graphical analysis of the hypothetical data suggests that $v = f\lambda$ for that medium.

Effects of Temperature on Sound

At a given temperature the speed of sound is constant in a given medium. Even though you may have available a solid, a liquid, and air at the same temperature, the speed of sound is constant in each medium but the speed has a different value in each medium.

Physical Harm

Like other phenomena that can be described as waves, sound waves diffract (bend around corners), refract (change direction when traveling from one medium to another), and reflect (bounce off surfaces).

Stand at a train station when a speeding train passes by with the horn blaring. When the train passed the station, the pitch of the horn seemed to suddenly change to a lower tone. That phenomenon is called the *Doppler effect*.

When the locomotive is coming toward you, your ears are receiving more waves each second than when the train is standing still. You are observing the speed of sound plus the speed of the train for the waves you observe with your ears. Similarly, the opposite change is observed when the train is moving away.

Reflection of Sound

The echo is a reflection of sound from a surface. Sound reflects from a surface very much like a ball bounces off a surface. Hearing your voice calling back from the bank of a distant canyon is an observation of an echo.

If the acoustics of a room are bad, reflections of sound from walls or other surfaces can meet in such a way as to cancel sound altogether. A sound wave is three-dimensional, spreading away from a source of sound in the shape of a sphere. If two sound waves meet at a point so that a rarefaction from one source meets a compression from another source of the same frequency, they will completely cancel each other.

Physical Harm _____

Sources of sound have images in reflecting surfaces in much the same way as images are formed by a mirror. Points where waves cancel are called nodal points. Two-dimensional waves like water waves cancel in lines called nodal lines. Waves in three dimensions cancel in planes called nodal planes. Nodal lines are in the shape of hyperbolas and nodal planes are in the shape of hyperboloids. Special conditions must exist for such a pattern to be formed.

Interference of Sound

The image of the tuning fork (about 10^3 cycles/s) held about 30 cm from a plane surface will have an image in the surface the same distance in back of the surface as the tuning fork is in front. Effectively we have two sources of the same frequency in phase and that sets the conditions for interference. The thing an observer can experience while walking away from the surface (from about 5 m from the source) is called an interference pattern.

All of these things occur because sound is a wave and the speed of sound is constant for a given temperature. The speed of sound in air is about 331.5 m/s at 0°C and increases about 0.60 m/s for every Celsius degree increase in temperature. In the FPS system, the speed is about 1,087 ft/s at 32°F and increases 1.1 ft/s for every Fahrenheit degree increase in temperature.

The Least You Need to Know

- A source of sound is an object vibrating at 20–20,000/s.

- Transverse waves are characterized by crests and troughs.

- Longitudinal waves are characterized by compressions or condensations and rarefactions.

- The speed of a wave is described by $v = f\lambda$.

Atoms, Ions, and Isotopes

In This Chapter

- ◆ Atoms
- ◆ Ions
- ◆ Isotopes

In this chapter, the atom and molecule are discussed as the building blocks of matter.

A Model of the Structure of the Atom

Imagine that you travel from your frame of reference into the insides of matter. The first particle you see is fairly large as these tiny particles go. It is called a *molecule*.

You may see molecules like hydrogen, water, or sugar. The hydrogen molecule is fairly small, a water molecule is a little larger, and as molecules go, a sugar molecule is huge. Prying a little deeper into matter you find tiny particles called *atoms*.

Atoms Next

Length for instance; the hydrogen atom is about $0.6 \overset{\circ}{A}$ in diameter.

Phun Phacts

It is important to be familiar with both the nanometer and the angstrom because both are used in reference books. 1 nanometer = 10^{-9} m; 1 angstrom = 10^{-10} m = 10^{-8} cm = 0.1 nanometer.

The largest atoms are around $5 \overset{\circ}{A}$ in diameter. $1 \overset{\circ}{A}$ or $1 \ \overset{\circ}{A}ngström = 1 \times 10^{-10} m = 1 \times 10^{-1} nanometers$. The mass of the hydrogen atom is about 1.673353×10^{-27} kg. We talk about these particles as if they have always been part of the daily experience. Scientists have developed most of the information known today about atoms since the latter part of the nineteenth century. Three models of the atom will be discussed briefly to provide a feeling about the evolving nature of how we think about atoms. Thomson associated a negative charge, an idea developed in a later chapter, with the electron, a constituent of the atom. He proposed a model that assumed there was an equal amount of positive charge associated with the atom, now known as protons. Bohr's idea about shells for the location of electrons was helpful, but his ideas came after Rutherford's discovery of the positively charged nucleus.

These two ideas improved Thomson's original model. Since electrons are attracted to the positive nucleus, there is a problem with the fact that electrons were not spiraling into the nucleus of the Rutherford model. Rutherford's contribution of the nucleus containing positive charge and most of the mass of the atom did, however, present an improved model over earlier ones.

The Bohr model used the nucleus of the Rutherford atom and proposed discrete orbits for the electrons to remedy the problem of having them spiral into the nucleus as they lost energy. His model provided for the electrons to jump into an orbit with higher energy by absorbing a quantum of energy. The electron gives up or emits a quantum of energy if it falls back into a more normal orbit. Scientists have continued to work on their interpretations of the particles of matter by improving the model of the atom.

Parts of the Atom

The nucleus of the atom is made up of *protons* and *neutrons*. The proton has a positive charge. In fact, the value of its charge has the same magnitude as the charge on the electron. The magnitude of the charge on the proton is 1.6×10^{-19} coulomb, and its mass is about 1.7×10^{-27} kg. The neutron is neutral, that is, it has no charge and a mass that is close to the value of the proton's mass.

The protons and neutrons in the nucleus are referred to collectively as *nucleons*. The number of neutrons and protons in the nucleus enable us to identify different kinds of atoms as well as to specify characteristics of a given atom.

We noted earlier that most of the atom is empty space. The atom does have the *electrons* outside the nucleus. Different models suggest that they are in orbits, shells, or clouds. In fact, there are just as many electrons outside the nucleus as there are protons in the nucleus. The electron is a particle with a mass of 9.1×10^{-31} kg and a charge of 1.6×10^{-19} coulomb.

The charge on the electron was established by Millikan's oil drop experiment, in which he also established the fact that the charge on any charged object was some integral multiple of the charge on the electron. That means that charge is grainy, with grains of definite size.

Physical Harm

Stating that the atom is mostly empty space emphasizes that the nucleus is only about 10^{-4} of the diameter of the atom. Think of it this way. Perhaps you have or had a dorm room. When you were in the room, you occupied a great deal more space than your body actually measures. In the same way, an atom takes up lots of space because of the electrons that are outside of the nucleus. The volume of the nucleus is only a part of the whole volume of the atom.

Isotopes and the Same Element

The model of the atom used here has two important numbers associated with it, the *atomic mass number* designated by the symbol A and the *atomic number* that identifies the number of protons in the nucleus, designated by the letter Z. These numbers are used to identify the element and to indicate the number of nucleons.

Suppose that you are given an atom with 4 protons and 12 nucleons. The atom has 12 nucleons made up of 4 protons and 8 neutrons. The symbol for an atom is written $^{12}_{4}Y$, where Y is the symbol of the element and can be found on a periodic table of elements. In general, the symbol for an atom is written $^{A}_{Z}Y$. As you can see, Z identifies the element even though the symbol for the element is included in the notation.

There are atoms of the same element that have a different number of neutrons. For example, all oxygen atoms have 8 protons, but they can have a different number of neutrons. $^{16}_{8}O$, $^{17}_{8}O$, and $^{18}_{8}O$ are all examples of atoms of oxygen. Nuclei having the same number of protons but a different number of neutrons are called *isotopes*.

The unit of measurement for nuclear mass is the *unified atomic mass unit* or u. Carbon twelve, $^{12}_{6}C$, was given the exact value of 12.000000 u on this scale. On the same scale, a proton has an atomic mass of 1.007276 u, a neutron has a mass of 1.008665 u, and a hydrogen atom has an atomic mass of 1.007825 u. That means that 1.66054×10^{-27} kg = $1u$.

The atomic mass number is the whole number closest to the atomic mass of an atom. That is, given that the atomic mass of oxygen is 15.9994 u, then the atomic mass number is 16. That means that this isotope of oxygen has 16 nucleons and since the atomic number of oxygen is 8, there are 8 protons and 8 neutrons in the nucleus.

Instead of working with the mass of one atom, many times the scientist deals with the mass of a *mole* of atoms. A mole of anything is Avogadro's number of the items.

Avogadro's number is 6.02×10^{23}. The mass of a mole of a substance turns out to be numerically equivalent to the atomic mass of the smallest particle of the substance. The mass of a mole of iron in grams is 56 g, assuming that the smallest particle of iron is one atom. You may check our previous answer by dividing the mass of a mole by Avogadro's number.

$\dfrac{.056 \, \text{kg}}{6.02 \times 10^{23}} = 9.3023 \times 10^{-26} \, \text{kg}$, which is essentially the same as before when we used not the atomic mass number but the atomic mass of iron.

The Formation of Ions

The electrons occupy regions of space called energy levels. Only a certain number of electrons can be in a particular energy level. All of the energy levels that have any electrons in them are filled for some atoms like argon, helium, krypton, and neon. These elements are stable; that means they do not combine with other elements. They are found on the far right side of the periodic table.

If an atom gives up a valence electron for some reason, since it has complete energy levels in its core, it may become a stable ion. It is no longer a neutral atom, but has become a positively charged particle called an *ion* because it now has one more proton than it has electrons. If an atom gains an electron for some reason, it may become a stable negatively charged particle because it has one more electron than protons. Neutral atoms become negative ions if they gain electrons to become stable ions.

Physical Harm

Even though an alpha particle is a positive ion, it quickly acquires two electrons from surrounding matter and becomes a neutral helium atom.

Another Source of Ions

An alpha particle is the nucleus of the helium atom. Helium has atomic number 2 and mass number 4. Given that information we know that the nucleus has 2 protons and 2 neutrons. Since an alpha particle ^4_2He is just the nucleus of the helium atom, it must have a positive charge of plus 2.

Some types of atoms are radioactive. That means that the nucleus spontaneously disintegrates and emits *alpha particles* ^4_2He, *beta particles* $^{\ 0}_{-1}e$, or *gamma rays*. Radium $^{226}_{88}Ra$ is a radioactive element that can cause cancer if the human body is exposed to it for long.

Uses of Radioactive Particles

Radium has been very useful in the treatment of cancer as well as a research tool. It is used in the laboratory as a source of alpha particles, among other things. Rutherford used radium in his experiment in 1911 when he discovered the nucleus of the atom. He used radium embedded in lead $^{207}_{82}Pb$. Since lead absorbs radioactive particles, he left a small opening through which alpha particles traveled to bombard very thin gold $^{197}_{79}Au$ foil. Rutherford and his assistants found that most of the alpha particles passed right through the gold foil, indicating that the foil is very porous to alpha particles. This meant that much of the atoms of gold must have been empty space. Otherwise the alpha particles would have been pushed from their paths by the charged particles within the gold atoms.

Much to his surprise, a significant number of the alpha particles were deflected at large angles to their original paths. Since the alpha particles travel at about 20,000 mi/s, that result suggested that the alpha particles were interacting with something charged positively, the same as his alpha particles. It also suggested that the positive charge of the gold atoms had to be in a relatively small volume. This lead to the idea of a positive nucleus for the atom.

Almost unbelievable to Rutherford was the result that some of the particles bounced straight back, opposite the direction they were traveling initially. That would be about as surprising as seeing a billiard ball bouncing straight back from a head-on collision with another billiard ball. Unlike the billiard balls that touch when they collide, the alpha particles must have collided with something very massive and charged positively. This confirmed his ideas suggested about the nucleus.

Rutherford called the thing that the alpha particles interacted with *the atomic nucleus*. Because the interactions were infrequent, he concluded that the nucleus was very small. The diameter of an atomic nucleus is about 10^{-14} m, which is the order of magnitude of the diameter of an electron.

Just think about it for a moment: an alpha particle, $_2^4He$, traveling at 20,000 mi/s, colliding with the nucleus of a gold, $_{79}^{197}Au$, atom! No wonder there was such a change in direction for the alpha particles. Of course Rutherford did not know their relative sizes then, but his experiment laid the groundwork for our level of understanding of the nucleus of the atom.

The Formation of Molecules

Alpha particles are not the only ions you know about. Some ions are very familiar to you, especially if you enjoy good food. Sodium, $_{11}^{23}Na$, has a single valence electron in its outer energy level that the sodium atom is just itching to give up. If the outer electron left, the sodium atom would then have eight electrons just like the noble gases found at the far right of the periodic table. Remember how stable they are. Chlorine is lacking one electron in its outer energy level and would be delighted to fill that shell and become more stable also. When sodium and chlorine atoms are brought into the vicinity of each other, sodium quickly gives up an electron to chlorine and becomes a stable positive ion while chlorine gains an electron becoming a stable negative ion. The attract each other and their combination is called sodium chloride.

Physical Harm _____

Electron transfer, electron sharing, and electron exchange are key ideas for remembering how molecules are formed.

It is called *ionic bonding*. In ionic bonding, one or more electrons are transferred from the outer energy level of one atom to the outer energy level of another atom. If the ions pull toward each other, the resulting ionic units are more stable than the original separated atoms. Since energy is given off during the electron transfer and ion attraction process the resulting product is more stable. This is like a ball rolling downhill. Once it is down, it is likely to stay there. It is energetically stable.

Other Ways of Forming Molecules

When two atoms share a pair of electrons to form a molecule, the combination is called *covalent bonding*. Atoms like hydrogen, chlorine, and oxygen exist in nature in pairs of atoms. The atoms establish a more stable particle by sharing electrons to complete an energy level. Because the first energy level is full if it contains only two electrons, two hydrogen atoms can share their electrons to complete the first energy level for each and become a more stable hydrogen molecule with symbol H_2.

Phun Phacts _____

Hydrogen and chlorine are two elements that occur in nature as molecules. Each molecule of these elements is made up of two atoms that share two electrons in a covalent bond.

Metals Exchange Electrons

The atoms of solid metals are very close together. They are so closely packed that electrons are shared freely throughout the atoms. The atoms of most metals are so closely packed that each atom of the element is surrounded by 12 other atoms.

The reason is that the electrons in the outer energy levels are loosely held and easily moved from one atom to the next. The sharing of electrons in this setting enables the atoms to attract each other. This type of bonding is called *metallic bonding*. These loosely held electrons that are continually exchanged are subject to being moved from one part of the solid to another by a special force.

The Least You Need to Know

◆ An atom is a neutral particle identified by its atomic number.

◆ An ion is a charged particle formed by an atom gaining or giving up electrons.

◆ Beta particles are emitted from some radio-active atoms.

13

Static Electricity

In This Chapter

- ◆ Static charge
- ◆ Insulators and conductors
- ◆ Detecting static charge

We now consider charged bodies and how they affect each other as well as the environment. We will charge an electroscope two different ways and see how it is used to detect a charge. There is no attempt in these discussions to count the number of charges involved.

Understanding Static Charge

Tear a few strips of newspaper about 2 to 3 inches wide and 15 to 20 inches long. Lay one of the strips on a rug, woolen shirt, or coat and stroke the strip of paper lightly with the back of your fingers. Lift the strip of newspaper, hold it up against the wall, and let go. Try holding another strip of newspaper against the wall that you have not stroked and let go. Place two strips of the newspaper on the rug and lightly stroke them with your fingers. Pick one of the strips up in each hand and bring the flat sides close together.

The two paper strips pushed away from each other. Pushed is the right word because as you try to move the strips closer together they move farther from each other. That is what static electricity will do, make your hair stand on end and curl your newspaper.

Scientists have found that there are two types of electrical charge of which static electricity is composed. One type is associated with a glass rod and the other type is associated with a hard rubber or amber rod. Rubbing the amber rod with fur enables you to cause that rod to be charged with one type of electrical charge. The glass rod rubbed with silk allows you to charge that rod with a different kind of electrical charge. When you walk across a carpet and open a door that has a metal doorknob, you often experience a jolt.

Benjamin Franklin named the two types of charge positive and negative. These are arbitrary names, that is, he could have named the charges Bill and Sue, but he chose positive and negative. These words do not have the same meaning as they do when you are talking about a number line. His theory proposed that when one type of charge is observed in one body, an equal amount of the opposite charge is observed in another body. The idea is that charge can move from one body to another, but charge is not lost. The algebraic symbols + and − are associated with these names to be consistent with the notion that the net charge created is zero. That is a crude statement of the conservation of charge. It does establish the important principle that ultimately no charge is lost or gained, or the total amount of charge in the universe is constant.

Insulators and Conductors

The strips of newspaper are an example of a type of matter that is called an *insulator*. Charge can be deposited on an insulator and the charges remain on the insulator where they are placed until they are wiped off for some reason. Insulators do not allow an easy movement of thermal energy. They also do not allow an easy movement of electrical charge.

The amber rod, the glass rod, and the hard rubber rod are all good insulators. All of these materials are great tools for studying static charge. Once charge is deposited on the insulator, it can be transported from one place to another. That charge is easily deposited by wiping an object to be charged with the charged insulator. The positive and negative charges are the same charges associated with the proton and electron, respectively.

Physical Harm

Electrons are free to move in a metal conductor while the protons remain in the relatively stationary nuclei of the atoms. Both positive and negative ions are free to move in a solution. Charge is a fourth fundamental quantity of physics considered so far.

The proton is fixed inside the nucleus of the atom and is not easily moved by ordinary means. That means that only the electrons are either taken off or placed on an insulator by friction by some other material like silk with glass during the charging process. When amber or hard rubber is rubbed with fur, electrons are deposited on the amber or hard rubber from the fur.

Conductors and Electrical Charge

Conductors are materials that allow a great deal of freedom of motion for the valence electrons; the electrons move freely throughout the material. Recall the metallic bonding in which electrons are attracted to several different nuclei at the same time, thereby making them fairly mobile rather than localized as is the case in other types of bonding. Those electrons can be pushed around the metal easily.

Charging an Insulator

The first law of static electricity is demonstrated by those strips: like charges repel each other, and unlike charges attract. That means that protons repel protons and attract electrons. In general, a body that has a deficiency of electrons is charged positively and an object with an excess of electrons is negatively charged. Two bodies attract each other if one is charged positively and the other has a negative charge.

Suppose that you charge a hard rubber rod negatively by rubbing it with fur. Friction helps you to deposit electrons on the insulator by rubbing them off the fur. The insulator is now negatively charged because it has an excess of electrons after being rubbed by fur. Now you may take the negatively charged insulator to the location of another insulator, rub some of the electrons off the first insulator onto the second, and then they both will be negatively charged. The second insulator has been charged by direct contact or just charged by contact.

Suppose that you rub the glass rod with silk so that friction enables you to remove some electrons from the rod and deposit them on the silk. The silk would be charged negatively because it would have an excess of electrons, and the glass rod would be charged positively because it would have a deficiency of electrons.

Physical Harm _____

Ground or zero potential energy has the same meaning for electrons as it does for objects when considering their gravitational potential energy with respect to the surface of the earth. Electrons at ground have zero electrical potential energy similar to how an object lying on the surface of the earth has zero gravitational potential energy.

Wire is a good conductor so the electrons move freely from the charged insulator through wire to your hand. Your body is a fair conductor and provides a path for the electrons to move to the earth or ground zero unless you are standing on well-insulated shoes. The earth is an infinite reservoir of electrons as well as an endless source of electrons.

Consider the results of an experiment in which it is found that 96,500 *coulombs* are carried by one mole of hydrogen ions. Remember the hydrogen ion is just one proton and is written symbolically, $_1^1H^+$. Hydrogen ions can be moved in a solution as can sodium ions and any other ion that goes into the solution. Recall that one mole is 6.02×10^{23} of anything. That means that each hydrogen ion has

$$\frac{96,500\,\text{coulomb}}{6.02 \times 10^{23}} = 1.6 \times 10^{-19} \text{ coulombs of charge.}$$

The charge on the proton is then $+1.6 \times 10^{-19}$ coulomb and since the charge on the electron has the same magnitude as the charge on the proton, the charge on the electron is -1.6×10^{-19} coulomb. Since the magnitude of the charge on the electron is 1.603×10^{-19} (to four significant digits) coulomb or $1.603 \times 10^{-19}\,\dfrac{\text{coulomb}}{\text{electron}}$, then there are

$$\frac{1\,\text{electron}}{1.603 \times 10^{-19}\,\text{coulomb}} = 6.250 \times 10^{18} \text{ electrons. That}$$

means that the unit of charge, one coulomb, is equivalent to 6.25×10^{18} electrons.

Detectors of Static Charge

The gold leaf electroscope or the aluminum foil electroscope is a very sensitive instrument for detecting electrical charge. It consists of a metal knob attached to one end of a metal rod that has two leaves attached to the other end.

The major components of the electroscope are
shown in Figure 13.1. The neutral electroscope is
shown in 13.1A, where the leaves hang limp with
equal positive and negative charges. The electroscope will indicate that a charge has been placed
on the knob and conducted to the rest of the conducting parts, including the leaves, when the leaves
spread apart.

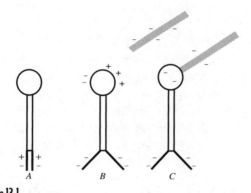

Figure 13.1

*The gold leaf electroscope is a sensitive detector of electric
charge.*

Another good detector of static charge is the *pith-
ball* electroscope. In such an electroscope, small
balls of pith are suspended on insulating strings.
Charge can be deposited on the pith-ball by touching it with a charged insulator. If an object of the
same charge is brought near the charged pith-ball,
it is repelled dramatically to some distance from the
charged object.

If an object with an unlike charge is brought near a charged pith-ball, the pith-ball swings dramatically toward the unlike charge.

Physical Harm

The pith-ball electroscope gives you an idea about the force that has been alluded to pushing the pith-balls apart. The force is very much like the gravitational force except its source is charge. The force is called the coulomb force and is calculated as $F = k \dfrac{Q_1 Q_2}{r^2}$, where Q_1 and Q_2 are the charges on the two bodies, r is the distance between the charges, and k is a constant that depends on the kind of insulator between the two charged bodies. Often the insulator is a vacuum or air.

Charge by Induction and Conduction

A procedure for charging the electroscope is summarized by the diagrams in Figure 13.1B and 13.1C. The hard rubber rod with a negative charge is brought in the vicinity of the knob of the electroscope in 13.1B. The electrons on the insulator repel the electrons on the metal knob and electrons move through the conductor away from the negatively charged hard rubber rod.

An excess of electrons remains in the leaves because if the insulator is taken away in Figure 13.1B, the leaves will hang limp again since the electrons will move back toward the knob rather than staying near each other in the leaves. The negatively charged insulator is allowed to touch the knob in Figure 13.1C, enabling electrons to actually move onto the metal knob. The electroscope is no longer neutral as a whole. It now actually has an excess of electrons. Refer to Figure 13.2 to see what happens when the insulator is taken away.

Figure 13.2

The negatively charged electroscope results from direct contact.

The knob collected some electrons when the insulator was in contact with it, leaving a net negative charge on the electroscope. If you bring a positively charged rod, one that has had electrons removed by friction, near the knob of the charged electroscope in 13.2, as indicated in the diagram of Figure 13.3A, the electrons move toward the knob.

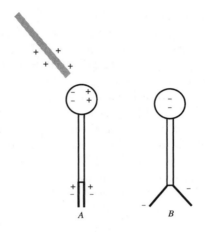

Figure 13.3

The electroscope detects charge if charge is in the vicinity of the metal knob of the electroscope.

They are attracted to the positive charge on the insulator. The electrons throughout the electroscope are attracted to the positive charge that is in the vicinity of the knob.

The leaves collapse, indicating that the net charge on the electroscope is apparently neutral. If the positively charged insulator is removed, as shown in the diagram of Figure 13.3B, the electrons flow back. The electrons move throughout the conductor, repelling each other until they are equally distributed. Then the leaves open to their original position, indicating that there is no net loss or gain of electrons due to the presence of the positive charge as long as it did not actually touch the electroscope.

The procedure outlined here is called charging the electroscope by direct contact. The charging body actually touches the electroscope.

Charging by Induction

Another method of charging the electroscope is charging by induction. An important principle is applied in this process to accomplish the charging of the electroscope. At the proper time, a path to or from ground is provided. The path to ground is the body of the experimenter. Simply touch the knob of the electroscope and electrons can either flow off to ground or travel from ground onto the knob. Refer to Figure 13.4 for a step-by-step procedure of charging by induction. The electroscope is shown in Figure 13.4A in its neutral state. The charging body is brought in the vicinity of the knob in Figure 13.4B causing the leaves to spread apart because there is an excess of electrons on them. The excess electrons are the ones that were repelled down from the knob.

When charging by induction, keep the source of charge far enough away from the electroscope so that it will not leave a residual charge on the knob by direct contact.

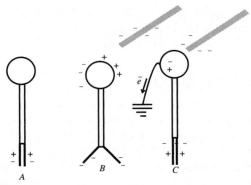

Figure 13.4

The first three steps in charging an electroscope by induction.

The ground connection is made in Figure 13.4C while the charging body is still in the vicinity of the knob. This means that you touch the knob of the electroscope with your finger. Excess electrons are repelled off the electroscope to ground in the direction of the arrow. Electrons are represented symbolically as $\bar{e}$. Enough electrons (only one is necessary) are repelled to ground to make the leaves of the electroscope appear neutral because they hang limp. Next the ground connection is removed from the knob in Figure 13.5A, while the charging body is still in the vicinity of the knob and the leaves of the electroscope appear to be neutral. The leaves are spread apart in Figure 13.5B, when the charging body is removed, showing that the electroscope does have a charge on it. The electrons distribute themselves evenly throughout the electroscope but there is now a deficiency of electrons.

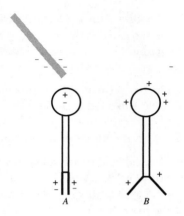

Figure 13.5

*The final steps in charging by induction show that the electro-
scope is a charge opposite the charging body.*

Since some electrons were repelled off to ground
and the ground was removed so that they could not
return, the electroscope has fewer electrons than it
had initially. The electroscope is left charged positively, the opposite charge of the charged insulator.
The charging body never lost any of its original
charge and never touched the electroscope during
this process.

Generating Much Larger Amounts of Charge

There are times that you might want to generate
a considerable amount of charge for some reason.
A *Van de Graaff generator* is used for that purpose
in the classroom. It is essentially a motorized version of the fur and insulator method of generating
electrostatic charge. The major difference is that

the charge is deposited on a spherical surface at the top of the generator. After the generator runs for a minute or so there is a tremendous amount of charge on the spherical surface.

When a conductor is isolated by an insulator, the charge placed on the conductor distributes itself on the surface of the conductor. None of the electrostatic charge is on the inside of the conductor.

A charged spherical surface has an *electrostatic field* associated with it, as diagramed in Figure 13.6, that is much like the gravitational field associated with the earth. One big difference is that there are two types of charge, so there are two different directions associated with electric fields. The direction is radially outward for positive charges and radially inward for negative charges.

If a conductor with a surface that is not spherical is charged the electrostatic charges will repel each other in an attempt to become distributed evenly on the surface. They become very crowded on parts of the surface that tend to be pointed. If there are points on the surface, the crowding of charge becomes so intense that the air molecules become ionized.

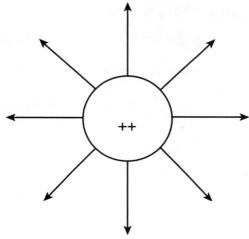

Figure 13.6

The direction of an electrostatic field is shown for a positively charged conductor.

Lightning and Charged Rain Clouds

Sharp objects like lightning rods can protect a house made of wood, bricks, or some other non-conducting material. The lightning rods can be placed at high strategic places on the building and, being well grounded, provide a path directly to ground for lightning. Any high object is a target for lightning. That is why a person should not take cover under a tree during a thunderstorm.

The Least You Need to Know

◆ A positively charged body has a deficiency of electrons.

◆ The electroscope is an excellent detector of charge.

◆ The coulomb is the practical unit of charge.

Chapter 14

Current Electricity

In This Chapter

- ◆ Current
- ◆ Potential difference
- ◆ Resistance

We are interested in controlled motion of charge so that we may use the energy of electrical charge in motion in our daily lives. The results of tamed electrons are used when we turn the lights on at home or turn the stereo on to listen to music. This chapter deals with the important components of simple direct current circuits and how they function to produce desired results.

Electrons in Motion

We have just completed a review of static electricity, but a brief look at electrons in motion was included in the last chapter. Any time you have electrons in motion that means there is a current. A flow of electrons means a current; references to a flow of current do not make a lot of sense.

Phun Phacts

Work done on a charge requires a force to be applied in a direction opposite to the coulomb force. That is similar to work done by the force lifting a body that acts opposite to the direction of the force of gravity on the body.

The object lifted to some height above the surface of the earth has potential energy with respect to the surface of the earth. The electrons deposited on the inside of the Leyden jar have electrical potential energy, or just electrical potential. They have potential because of the work done on each of them to force them onto the negatively charged plate of that capacitor.

Remember like charges repel each other with a coulomb force given by $F = k\dfrac{Q_1 Q_2}{r^2}$, where Q_1 and Q_2 are charges in coulombs, r is the distance between the charges, and k is a constant that is a property of the medium separating the charges. The units of k are $\dfrac{N - m^2}{\text{coulomb}^2}$ so that F is in newtons. The value of k for air is $9 \times 10^9 \dfrac{N - m^2}{\text{coulomb}^2}$. Notice that the force is positive if the charges are either both positive or both negative.

We discussed what happens when work is done on electrons and gives them a high potential. The electrons on the negative plate of the Leyden jar are so pumped up that if you should point at the jar, they can give you a bad-hair day and curl the leaves of your electroscope! Electrons in motion, electric current, can do work as they return to a lower potential. Electrical potential similar to potential energy can be transferred into other forms of energy like sound, heat, and light.

Electric Current

We observed the effects of electrons in motion; the electric current in lightning rods provide a path for the charges in a bolt of lightning to go straight to ground.

The lightning rods are conductors that contain lots of electrons that are exchanged by atoms in the conductor. The electrons are free to move from one end of the conductor to the other. The current we are considering at this time is more like the electrons from the lightning rod to the ground through a metal conductor. The conductor does not have to be metal; it can be ionized air like that which enabled the discharging of the Leyden jar. But the emphasis at this time is the continuous flow of charge through metal conductors to do work on something in its path.

Electric current is the rate of flow of charge or the flow of charge per unit of time. In symbols, $I = \dfrac{Q}{t}$, where I is current, Q is the charge in coulombs, and t is time. If the charge is 1 coulomb, 6.25×10^{18} electrons, and the time is 1 second, then the current is defined to be 1 ampere. The ampere, $I = \frac{1\,C}{1\,s} = 1\,A$, is the practical unit of electric current.

The ampere is a very large unit, so more often than not we use the milliampere, mA, or the microampere, μA, as a unit of current.

Potential Difference

Since charging the Leyden jar was a major task, it would be nice to have a good source of continuous current without the worry of charging a capacitor all the time.

Physical Harm

A small amount of current can damage the human body. Around 50 mA causes considerable pain. Around 100 mA can cause problems with the heart and even stop an unhealthy heart. Anyone who tells you that direct current will not hurt you needs to know about the Leyden jar, the wrong side of a television picture tube, and a lightning bolt.

We use several sources of continuous current in our everyday lives: the C-cell, the D-cell, or the storage cells in the storage battery of cars and trucks. These cells do internal chemical work on electrons to provide them with potential energy. Think of one pole of the battery as having electrons at high potential and the other pole as being a place of low potential. Refer to the cell as your source of continuous current with a certain potential difference associated with it.

Since the *potential difference* is the work done per unit of charge, we may find the units of measurement for potential difference. First, the symbol for potential difference is V, and the definition for V in symbols is $V = \dfrac{W_k}{Q}$. The units of V are joules/coulomb. If the work done is 1 joule and the charge moved is 1 coulomb, the potential difference is 1 volt, or 1 V = 1 joule/1 coulomb. The instrument that is used to measure potential difference in a circuit is called a *voltmeter*.

A battery may be needed to provide a potential difference of 3.0 V or 4.5 V. A special battery may be needed to provide twice as much current as we get from a single cell. The current referred to here is the continuous current needed to power a flashlight or a transistor radio.

Resistance

There is always some production of thermal energy associated with resistance in an electric circuit. This means that some of the electrical potential energy is converted to thermal energy. Sometimes sound is associated with resistance in a circuit. If the *resistor* is a filament in a lamp, there is some light in addition to much thermal energy associated with resistance.

The Unit of Resistance

Resistance is measured in ohms, denoted by the symbol Ω. Given a fixed potential difference in a simple circuit, the greater the resistance in ohms the smaller the current in amperes. The instrument used for measuring resistance is the ohmmeter.

Components of an Electric Circuit

The major features of an electric circuit are connecting conductors, resistance, potential difference, and current. Refer to Figure 14.1 to see how all of these components form a circuit. The diagram in Figure 14.1 is a schematic diagram of a basic electric circuit. The break in the circuit is the symbol for a switch. No current is shown in the circuit because the switch is open or in the off position.

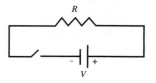

Figure 14.1

A simple electric circuit has a switch, a cell, and a load.

Phun Phacts

Use the switch in an electric circuit. Close it when the circuit is to work and open it when current in the components is not desired. A switch can save components and measuring instruments when a connection may be in error.

The current passes through a resistor represented by a squiggly line and labeled R in the diagram, and of course through the switch when it is closed. When the switch is opened, the current stops; the switch places the operator in control. The symbol for current is I. The current is not included in Figure 14.1 to emphasize that the switch is open.

An electric circuit has at least one resistor of some sort, usually referred to as the load, symbol R_L, or total resistance, R_T. It must have a switch. It also has connecting conductors, symbolized by straight lines between components in Figure 14.1, and a source of potential difference that provides a continuous current.

Series Circuits

The electric circuit in Figure 14.2 is called a series circuit because the resistors are connected end to end like links in a chain.

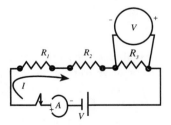

Figure 14.2

A series circuit contains resistors connected in a chain end to end so that the total current passes through each of them.

Resistors connected in a circuit in this way will all have the same amount of current through them. In Figure 14.2, the total current, I, in the circuit passes through each of the resistors, labeled R_1, R_2, and R_3. The ammeter is connected in series with the load, along with the voltmeter that is connected to measure the potential difference across R_3. Notice that the voltmeter is connected with its negative side closest to the negative side of the cell and its positive side closest to the positive side of the cell; it is said to be connected in parallel.

Physical Harm _____

The positive and negative sides of the electrical meters must be connected properly to the circuit not only so that they will measure correctly but also to protect the meter. For example, the voltmeter is wired internally so that most of the current stays in the circuit and only a small amount passes through a voltmeter properly connected in the circuit.

The ammeter is labeled the same way and is connected in series positive to positive and negative to negative. The meters are labeled in this way so that the DC current passes through the meter in the proper direction. Note well that the ammeter is not connected in the circuit in the same way as the voltmeter. The electrons give up energy as they travel around the circuit. For that reason, the decrease in potential energy or decrease of potential is called a potential drop.

Calculate the readings on the voltmeter and the ammeter in Figure 14.2. The total current is in each of the resistors in Figure 14.2 because they are connected end to end, that is, in series—the resistance of the load is the sum of the values of the resistors. The resistance of the load, or the total resistance, of any number of resistors connected in series is the sum of all of the resistors. Suppose $V = 1.5V$, $R_1 = 10.0\Omega$, $R_2 = 20.0\Omega$, and $R_3 = 30.0\Omega$; calculate I, V_1, V_2, and V_3.

$$R_L = R_1 + R_2 + R_3$$
$$V = IR_L$$
$$I = \frac{V}{R_L} = \frac{1.5V}{60.0\Omega} = 0.025A \text{ by Ohm's law}$$
$$1V = (1A)(1\Omega).$$

That means the reading on the ammeter is 0.025 A. Since $I = 0.025$ A, $V_1 = IR_1 = (0.025A)(10.0\Omega) = 0.25V$, $V_2 = IR_2 = (0.025A)(20.0\Omega) = 0.50V$, and $V_3 = IR_3 = (0.025A)(30.0\Omega) = 0.75V$.

We check to see if we are ready to go, by using Kirchhoff's second law: $V_1 + V_2 + V_3 = 0.25V + 0.50V + 0.75V = 1.5V$, and that is V!

Take a look at Figure 14.3, where you find a diagram of three resistors connected in parallel. Notice that the current is not the same in each of the resistors but each *branch* has its own current.

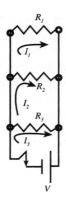

Figure 14.3

The electric circuit with three resistors connected in parallel demonstrates branch currents.

Parallel Circuits

Kirchhoff's first law is the third rule for this process, and it states that the total current passing though a part of a circuit in which resistors are connected in parallel is equal to the sum of the currents passing through each branch of the parallel part of the circuit. That is essentially a statement of conservation of charge; no charge is gained or lost when circuits branch into different paths. Kirchhoff's second law is a statement of the conservation of energy—that is, no energy is gained or lost in electric circuits. The symbolic statement of that rule for Figure 14.3 is $I = I_1 + I_2 + I_3$. Suppose the resistors and the potential difference all have the same values as in the last circuit. Calculate the total current and the branch currents in the diagram of Figure 14.3.

Notice that the potential drop across all resistors in this circuit is the same because each has one end connected to the negative side of the cell and each has the other end connected to the positive side of the cell. That means that the potential drop across each resistor is V and is stated symbolically as $V_1 = V_2 = V_3 = V$, $I = I_1 + I_2 + I_3$,

$$\frac{V}{R_L} = \frac{V_1}{R_1} + \frac{V_2}{R_2} + \frac{V_3}{R_3}, \ \frac{1}{R_L} = \frac{1}{R_1} + \frac{1}{R_2} + \frac{1}{R_3} \text{ (dividing}$$

both members of the previous equation by V *the potential of the cell).*

The terms in the right member are added by fist finding a common denominator, which is $R_1 R_2 R_3$ for this circuit: $R_L = \dfrac{R_1 R_2 R_3}{R_1 R_2 + R_1 R_3 + R_2 R_3}$.

The resistance of the load in Figure 14.3 is

$$R_L = \frac{6000.0\Omega}{1100.0} = 5.45\Omega.$$

$$I = \frac{V}{R_L} = \frac{1.5V}{5.45\Omega} = 0.28A, \ I_1 = \frac{1.5V}{10.0\Omega} = 0.15A,$$

$$I_2 = \frac{1.5V}{20.0\Omega} = 0.075A, \text{ and } I_3 = \frac{1.5V}{30.0\Omega} = 0.050A.$$

Use Kirchhoff's first law to check your result: $I = 0.15A + 0.075A + 0.050A = 0.28A$.

Series-Parallel Circuit

Use the same components that we started out with except the potential difference of the cell has the same number of significant figures as the resistors, that is, $V = 1.50V$. Connect the resistors in a series-parallel circuit. The schematic diagram in Figure 14.4 is one arrangement.

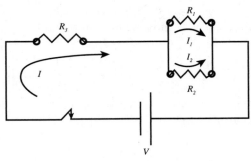

Figure 14.4

Three resistors are connected in a series-parallel circuit.

$$\frac{1}{R_{eq}} = \frac{1}{R_1} + \frac{1}{R_2}$$

$$R_{eq} = \frac{(10.0\Omega)(20.0\Omega)}{30.0\Omega} = 6.67\Omega$$

$$I = \frac{V}{R_L} = \frac{1.50v}{36.7\Omega} = 0.0409A$$

$$V_3 = IR_3 = (0.0409A)(30.0\Omega) = 1.23V$$

$$V_1 = V_2 = V - V_3 = 1.50V - 1.23V = 0.27V$$

$$I_1 = \frac{V_1}{R_1} = \frac{0.27V}{10.0\Omega} = 0.027A$$

$$I_2 = \frac{V_2}{R_2} = \frac{0.27V}{20.0\Omega} = .014A$$

That completes the solution, and now to check using Kirchhoff's first law:
$I = I_1 + I_2 = 0.027A + 0.014A = 0.041A$.

Suppose you need a battery that supplies 3.0 V and 0.25 A. Refer to Figure 14.5 to see the 3.0 V battery needed.

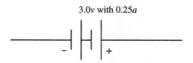

Figure 14.5

A 3.0 V battery made of cells delivers the same current as one cell.

The cell is like a resistor in that the potential differences add when connected in series. The currents add when cells or batteries are connected in parallel. In order to construct a 3.0 V battery, you simply connect two cells in series. The cells shown in Figure 14.5 are connected in series. Suppose you need a 3.0 V battery that supplies 1.0 A of continuous current and you are to construct it from the same supply of cells. Figure 14.6 is a schematic of the 3.0 V battery that supplies 1.0 A of current.

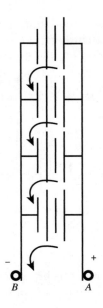

Figure 14.6

A battery that supplies current contains cells or batteries connected in parallel.

Notice that two cells are connected in series to have a 3.0 V battery and then four 3.0 V batteries are connected in parallel to have 1.0 A of current since each branch contributes 0.25 A. Because the batteries are connected in parallel, the potential difference from *A* to *B* is 3.0 V. That means the anodes of all the 3.0 V batteries are connected together and all the cathodes of those batteries are connected together. The anode of the 3.0 V and 1.0 A battery is *A* and the cathode is at *B*.

If the resistor has a power rating that is too small, the resistor will quickly burn out and act like a switch and open the part of the circuit in which it is connected. We have the background necessary to figure out the part played by the power of a resistor.

$P = \dfrac{W_k}{t}$ and $V = \dfrac{W_k}{Q}$ (so work equals voltage

times charge), $P = \dfrac{VQ}{t} = \dfrac{Q}{t}V = IV$, and

$P = I(IR) = I^2R$. You have discovered with that analysis that power can be calculated by finding the product of current and potential difference or by squaring the current and multiplying by the resistance.

The Least You Need to Know

- ◆ A simple circuit requires a switch, source of potential difference, load, and connecting conductors.

- ◆ The current in a simple circuit may be calculated with Ohm's law, $V = IR$.

- ◆ The total resistance of resistors connected in series is the sum of the individual resistors.

- ◆ The equivalent resistance of resistors connected in parallel is the reciprocal of the sum of the reciprocals of the individual resistors.

15

The Particle Model of Light

In This Chapter

- ◆ Understanding the particle model
- ◆ Direction and speed of light particles
- ◆ Reflection and refraction of light

The behavior of light is explained in this chapter by the use of models. Two models are considered here, beginning with the particle model.

The Particle Model

If light is a particle, it must have a source such as the sun, the glow of the filament of a light bulb, or the flame of a candle. The particles must be very small to be perceived by the eye. The particles travel in straight lines radially outward from a point source because everyone in a dark room sees a source at the same time. The particles travel at a high speed, in fact, the fastest speed ever observed.

Physical Harm

Light does travel at a constant speed, but it does not travel at the same speed in all media. For example, light travels at a speed in water that is about three fourths of the speed of light in a vacuum. The speed of light in air is slightly less than the speed in a vacuum.

The Inverse Square Law

Suppose you have a point source of light with particles traveling radially outward from that point as diagrammed in Figure 15.1.

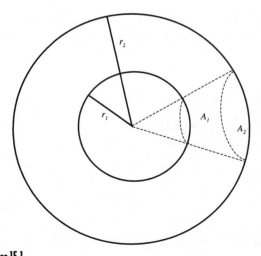

Figure 15.1

The illuminated surfaces are parts of two concentric spheres.

Imagine that we could have all the particles (N) from the point pass through two concentric spheres. A cutaway diagram of those spheres, one with radius r_1 and the other one with radius r_2, has the point source at their center.

$$\frac{I_1}{I_2} = \frac{\dfrac{N}{A_1}}{\dfrac{N}{A_2}} = \frac{A_2}{A_1} = \frac{4\pi r_2^2}{4\pi r_1^2} = \frac{r_2^2}{r_1^2}.$$

In general, the intensity of illumination, I, is inversely proportional to the square of the distance from the source. That is, $I = \dfrac{k}{r^2}$, where k is the constant of proportionality that turns out to be the luminous intensity of the source of light in candela.

Direction and Speed

The paths of particular particles are called rays, and a bundle of rays is called a pencil of light or a beam of light. The particles are so small that two beams or pencils of light can cross and the particles never deviate from their paths or bounce off each other.

Phun Phacts

Light is different than sound in a lot of ways, but one huge difference is that light does not need a medium for transmission.

The motion of light is described as uniform motion. The distance to the sun from the earth is about

92,000,000 miles. How long does it take for light to travel from the sun to the earth?

$$d = 92,000,000 \text{ mi}, v = 186,000 \text{ mi/s}, t = ?$$

$$d = vt, \quad t = \frac{d}{v}$$

$$t = \frac{92,000,000 \text{ mi}}{186,000 \text{ mi/s}}$$

$$t = 490 \text{ s} = 8.2 \text{ min.}$$

If the sun suddenly stopped radiating light, we would not know it for a little over eight minutes!

Reflection

Light is found to have two laws that it obeys when it reflects from a smooth surface:

- The incident ray, the reflected ray, and the normal to the surface all lie in the same plane.
- The angle of incidence is equal to the angle of reflection.

For our discussion, a plane mirror will serve as an adequately smooth surface. Refer to Figure 15.2 to see how a plane mirror forms an image.

Images in a plane mirror are located by finding the point of intersection of two virtual rays. The virtual rays appear to come from the image along the same line of sight as the reflected ray. A vertical arrow labeled O in the diagram represents the object and the image is labeled I.

Phun Phacts

A real image is formed when real reflected or refracted rays of light actually intersect. A virtual image is formed when virtual extensions of reflected or refracted rays of light only appear to intersect.

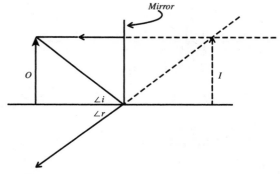

Figure 15.2

The image formed by a plane mirror is located by reflected rays.

The image formed by a plane mirror is described qualitatively being as far behind the mirror as the object is in front—it is a virtual image, it is right side up, and it is the same length as the object.

Reflection and Spherical Mirrors

Light obeys the same laws of reflection any time it is incident on a smooth surface. Even if the surface is curved, the same laws enable you to locate the image formed by a curved mirror. Suppose that a mirror is cut from a portion of a hollow metal sphere in which the inside surface is polished. That mirror is called a *concave spherical mirror*. If the outside surface is polished to reflect, it is called a *convex spherical mirror*.

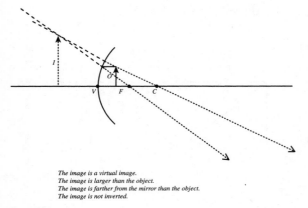

The image is a virtual image.
The image is larger than the object.
The image is farther from the mirror than the object.
The image is not inverted.

Figure 15.3

The image formed by a concave spherical mirror when the object is between V and F.

The mirror has a special point associated with it: the center of the sphere that the mirror came from. The point is called the *center of curvature, C.* The principal axis is a line that is associated with the mirror and is one that passes through the center of curvature while intersecting the mirror perpendicularly at its center, the vertex, *V.*

Quantitative Descriptions of Images

A quantitative description of the image can be made as well. That involves actual measurements of distances such as f, S_o, and S_i. A relationship is derived using Figure 15.4. The relationship works for all cases in which an image is formed by a concave spherical mirror.

The distance from F to the object is labeled S_o and the distance from F to the image is labeled S_i. Both distances are positive if measured from F away from the mirror and negative if measured from F toward the mirror.

The height of the object is labeled H_o and the height of the image is H_i in the diagram. The ratio $\dfrac{H_o}{H_i}$ enables you to find the magnification of the image formed by a spherical mirror.

$\dfrac{f}{S_i} = \dfrac{S_i - f}{f - S_o}$ by the transitive property of equality.

Therefore, $S_o S_i = f^2$.

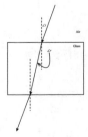

Figure 15.4

The quantitative description of the image formed by a concave spherical mirror applies to all cases when an image is formed.

Refraction

Trace the path of a ray of light as it travels from air into glass and is transmitted through a rectangular glass prism as shown in Figure 15.5.

It is obvious that the angle of refraction in glass is smaller than the angle of incidence in air even without measuring the angles.

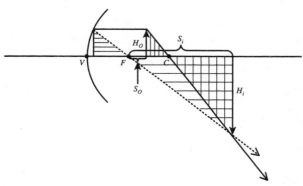

Figure 15.5

A light ray passing from air into glass bends toward the normal.

Mathematics of Snell's Law

The simple relationship that replaced many volumes of books containing observations of angles of incidence and corresponding angles of refraction is called Snell's law and is written $\dfrac{\sin \angle i}{\sin \angle r} = n$, where n is a constant called the index of refraction and is associated with the type of *medium* the light is entering, assuming that it is leaving a vacuum.

The index of refraction of a vacuum is defined as 1, and for air it is so close to that value that we use 1. The index of refraction of glass is $\frac{3}{2} = 1.50$ and the index of water is $\frac{4}{3} = 1.33$.

Check the behavior of the ray of light entering glass in Figure 15.5 by measuring the angle of incidence in air and using Snell's law to calculate the corresponding angle of refraction. The angle of incidence is about 20° so by Snell's law the angle of refraction is ...

$$\sin \angle r = \frac{\sin 20°}{1.5}$$
$$\sin \angle r = 0.2280$$

When the value is calculated, the answer is about 13°.

Physical Harm

If a ray of light is traced from a medium that has an index of refraction greater than the index of refraction of the medium the ray is entering, then there will always be a critical angle. There will never be a critical angle going from a medium whose index of refraction is less than the index of refraction of the medium it is entering, such as going from air into glass.

The process of a light ray leaving the prism is just the inverse of the process of a light ray entering the prism so that Snell's law states …

$$\frac{\sin \angle i_{glass}}{\sin \angle r_{air}} = \frac{1}{n_g} \text{ (the reciprocal or inverse of } n_g\text{)},$$

$$\angle r_{air} = 20°.$$

The Critical Angle of Incidence

The last ray that will leave glass and enter air leaves at an angle of 90° in air. The angle in glass that would cause that angle of refraction is called the critical angle for glass That is, Snell's law states $\dfrac{\sin \angle i_{glass\ crit}}{\sin 90°} = \dfrac{1}{n_g}$ or

$$\sin \angle i_{glass\ crit} = \frac{1}{n_g} \text{ since the value of } \sin 90° \text{ is } 1.$$

That means that $\angle i_{glass\ crit} = \sin^{-1}\left(\dfrac{1}{n_g}\right)$ or

$$\angle i_{glass\ crit} = \sin^{-1}\left(\frac{1}{1.5}\right) = 41.8°.$$

That means that any angle of incidence in glass less than or equal to 41.8° will leave the glass and enter air. Any ray at an angle of incidence in glass greater than 41.8° will be totally reflected back into glass. That is called total internal reflection.

Snell's Law for Two Media Other Than Air

Suppose we have two prisms with parallel sides separated by a thin layer of air. Trace a ray of light from air through the prisms and the layer of air as shown in Figure 15.6.

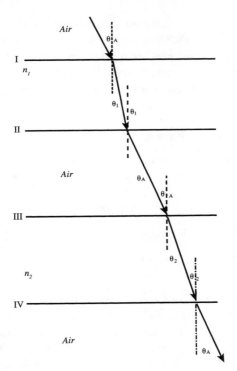

Figure 15.6

A ray of light travels from medium 1 into medium 2 according to the expression $n_1 \sin\theta_1 = n_2 \sin\theta_2$.

Start with the surface at I and end at surface IV.

$$\frac{\sin\theta_A}{\sin\theta_1} = n_1 \text{ , at I}$$

$$\frac{\sin\theta_1}{\sin\theta_A} = \frac{1}{n_1} \text{ , at II and similar statements can be}$$

made for the remaining surfaces.

Both the first and second equations show that $\sin \theta_A = n_1 \sin \theta_1$ and both of the third and fourth equations show that $\sin \theta_A = n_2 \sin \theta_2$. If you look at Figure 15.6, you see that all of the normals are parallel and that means all of the angles in air, θ_A, are equal.

> **Phun Phacts**
>
> Since the normals are all parallel because the faces of the prism are parallel, the angles in each medium are equal. The reason is that when parallel lines are cut by a transversal, the ray of light here, the alternate interior angles are equal in measure.

If the layer of air becomes smaller and smaller, then the medium with index of refraction n_1 is not separated from the medium with index of refraction n_2. Therefore the expression, $n_1 \sin \theta_1 = n_2 \sin \theta_2$, enables you to describe quantitatively the behavior of light when it travels any transparent medium into any other transparent medium.

The ray diagram in Figure 15.7 shows how a single light ray A is partially reflected as B and partially refracted as F. The refracted portion is incident on the bottom of the glass prism and is partially reflected as G and partially refracted into air as C.

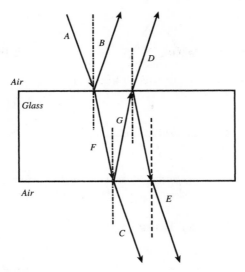

Figure 15.7

Reflection and refraction at the same surface at both parallel faces of a glass prism.

The ray *G* is incident on the upper surface of the glass prism where it is partially reflected back toward the bottom and partially refracted into air as *D*. The ray *D* emerging into air is parallel to the reflected ray *B*, and *E* emerges into air traveling parallel to *C*. Trace a ray of light through at least one other shape of prism. The glass prism in Figure 15.8 provides the opportunity to do just that. The ray *A* incident on the prism initially is parallel to the base of the prism. The ray *B* that emerges into air is obviously deviated from the original direction of *A* to a direction toward the base of the prism.

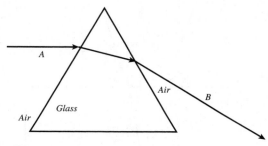

Figure 15.8

Tracing a light ray passing through a triangular prism of glass shows it emerging closer to the thicker portion of the prism.

Physical Harm

An incident light ray that is parallel to one side of a glass triangular prism passing through the thin portion of the prism emerges into the second medium traveling more toward the thick portion of the prism.

The Double Convex Lens or Converging Lens

Suppose you make a new prism with nearly parallel sides in the middle and triangular-shaped edges. A polished prism like that is called a converging or double convex lens. That is because it converges rays of parallel light into a focus area. The lens in Figure 15.9 is such a prism and uses those features to form images.

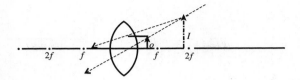

The image is between f and $2f$.
The image is a virtual image.
The image is larger than the object.
The image is not inverted.

Figure 15.9

The image formed by a double convex lens can be a virtual image.

Each side of the lens is convex so the lens does not have one center of curvature like the concave mirror. Distances from the middle of the lens are measured in terms of the focal length, f, and twice the focal length, $2f$, where f is the distance to the principal focus, F, on both sides of the lens.

The images are formed where the refracted rays intersect or appear to intersect. The image of only one point is located because images of all other points are found the same way. The image of the base of the object is located on the principal axis if the object's base was on the principal axis.

The quantitative description of the images formed by a double convex or converging lens is outlined using the image in Figure 15.10.

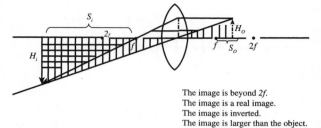

The image is beyond *2f*.
The image is a real image.
The image is inverted.
The image is larger than the object.

Figure 15.10

The diagram enables you to develop the quantitative description of an image formed by a double convex lens.

Quantitative Description of Images Formed by a Lens

The distance to the image, S_i, is measured from the principal focus, on the side of the lens opposite the object, to the image. That distance is positive if it is measured from the principal focus moving away from the lens. The distance to the object, S_o, is measured from the principal focus to the object on the same side of the lens as the object. That distance is positive if it is measured from the principal focus moving away from the lens. All distances are measured along the *principal axis*.

The right triangles that are marked with vertical lines are similar.

$$\frac{H_o}{H_i} = \frac{f + S_o}{f + S_i}$$ Corresponding sides of similar triangles are proportional.

$$\frac{H_o}{H_i} = \frac{f}{S_i}$$ Corresponding sides of similar

triangles are proportional.

$$f^2 = S_o S_i .$$

Testing the Particle Model

How can particles know to change directions when traveling from one medium to another? Since the problem seems to be refraction, an experiment can provide an answer. The experiment involves treating light as a particle modeled by a particle that is familiar and checking to see if Snell's law is obeyed by the particle. The particle is a large steel ball.

The media are modeled by two horizontal surfaces separated by an inclined plane about a textbook thickness high. The particle is launched at the same speed on the upper surface and is allowed to roll down the inclined plane to the lower surface. The path of the ball on both surfaces is recorded by carbon paper tracks on plain white paper. The angles of incidence are matched up with the corresponding angles of refraction.

Snell's law does explain the change in direction of the ball on the lower surface compared to the upper surface. The angles on the lower surface were smaller than the angles on the upper surface so the upper surface is a model of air and the lower surface water or glass. There is a problem here even though the change in direction can be explained. If the ball rolls down the inclined plane it travels faster on the lower surface than it does on the upper surface. That just does not agree with our experience because we know that light travels faster in air than it does in water.

The particle model for light does not do a complete job of explaining the behavior of light. The first inclination is to toss it out because it cannot explain refraction. But look at all of the things it does help to explain. The logical thing to do is to continue using the particle model to explain things that it helps us explain. Then we can look for a better model that can handle everything that the particle model makes clear as well as things the particle model cannot explain. We begin the search for a more comprehensive model for light in the next chapter.

The Least You Need To Know

- ◆ Use a ray diagram to locate the image of an object in a plane mirror.

- ◆ Use a ray diagram to locate an image formed by a concave spherical mirror and describe it both qualitatively and quantitatively.

- ◆ Draw the ray diagram and describe the image of an object formed by a double convex lens.

Chapter 16

Light and Waves

In This Chapter

- ◆ Reflection of waves
- ◆ Refraction of waves
- ◆ Diffraction of waves

The wave is now considered as a possible model for light. We consider waves in springs and strings as well as waves in water in this chapter.

Reflection

Two springs and some strong string are good tools to study waves. Lay the spring on the floor and shake a transverse pulse into the spring for observation.

The two pulses shown in Figure 16.1 traveling toward the anchored end in *IA* were launched rapidly, one after the other.

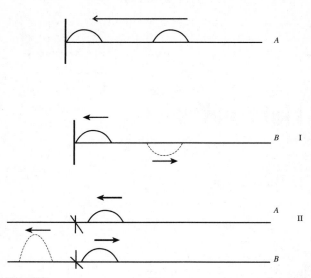

Figure 16.1

The waves reflected at different junctions are reflected differently.

The first pulse reflected off the anchored end in *IB* and the second pulse is approaching that same junction. The reflection is very much like the reflection of light with the added information that the reflection is inverted when the *junction* is an anchored end.

Behavior of a Wave at a Junction of Two Media

Figure 16.1 shows a pulse in II*A* traveling toward the junction that is marked with an *X*. The pulse is partially reflected from the junction and partially transmitted across the junction as shown in II*B*. In addition, the reflected pulse is larger and on the same side of the spring as the original pulse. The part of the original pulse that was transmitted across the junction was larger and traveled much faster than the original pulse!

The string was tested by itself and it was found that a pulse travels several times as fast in the string as in either spring with the same tension. The two springs were connected together to check to see if the same thing happened as with the string. A pulse traveling from the large diameter spring to the smaller diameter spring reflected upside down at the junction and part of the pulse was transmitted into the smaller diameter spring right side up, smaller, and at a slower speed.

Transverse Waves That Superpose

Look at a chain of periodic waves. The waves discussed here are transverse with a frequency and wavelength established by the source generating the waves. Keep in mind that the series of waves is traveling at a constant speed and we are able to observe only some of the waves. In Figure 16.2, there is a sequence of diagrams where the waves move a quarter of a wavelength at a time for the purpose of discussion.

The incident chain of waves continues to move until three fourths of a wavelength is reflected, as shown in Figure 16.2(D). At that instant, the medium would be horizontal in that region because the pulses superpose yielding a sum of zero displacement. In Figure 16.2(E), a full wavelength is reflected and superposes with a full wavelength of the incident wave.

Physical Harm

The waves reflected from the barrier appear to be coming from a virtual image of the source on the other side of the junction. The reflected waves have the same frequency as the source.

Neither individual pulse can be seen because the waves superpose, giving a sum that is one larger pulse in both regions of the incident wave. The chain of pulses continues to move to the junction superposing with a chain of pulses reflected from the junction until the pattern, shown in Figure 16.2(F), is seen. Regions called *loops* or *anti-nodes* that are half a wavelength apart as shown, separated by *nodes* or *nodal points*, characterize the total pattern. The pattern is called a *standing wave*. The wavelength of the standing wave is one half the wavelength of the source.

Calculating Speed from a Standing Wave Pattern

If you determine the wavelength of a standing wave and the frequency of the source, you may calculate the speed of the wave in the medium using the relationship we developed earlier for periodic waves.

Physical Harm

Nodal lines are made up of all the points of destructive interference generated when waves from two point sources pass through each other.

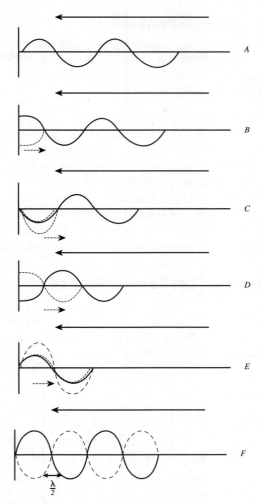

Figure 16.2

The waves reflected from a fixed junction superpose to form a standing wave.

Waves in Two Dimensions

A good way to begin consideration of a two-dimensional wave is to use a ripple tank. A *ripple tank* is a piece of laboratory equipment used to study two-dimensional waves.

A fingertip at the water surface serves as the source of waves that travel radially outward at a constant speed in all directions in a plane. The speed is constant because the wave remains a circle until it strikes the sides of the tank. A plastic ruler or a wooden dowel held lengthwise may be used to touch the surface of the water to generate a straight wave. The speed is constant because the wave remains straight until it strikes the sides of the tank.

Reflection of Two-Dimensional Waves

A barrier placed in the tank serves the same purpose as anchoring the spring to the floor. The straight waves striking the barrier reflect as shown in Figure 16.3.

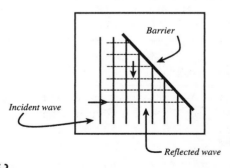

Figure 16.3
Straight waves reflect from a barrier in a ripple tank.

The reflection of the straight waves is a little different than the reflection of rays of light, but remember this is a two-dimensional wave. Dark lines in Figure 16.3 represent the crests of the incident waves. The corresponding crests of reflected waves are represented by dotted lines. Arrows perpendicular to a crest of each indicates the direction of propagation of each set of waves. A closer look shows that the angle of incidence is about equal to the angle of reflection. The angle of incidence is the smaller angle between the incident wave and the barrier.

The angle of reflection is the smaller angle between the reflected wave and the barrier. If the path of a small segment of each wave is traced, the paths behave very much like rays of light. The straight waves reflecting from the barrier are very similar to the light rays reflected from a plane mirror.

Waves Reflected from a Curved Mirror

A model of a curved mirror for the ripple tank is challenging. Either a piece of flexible metal or a piece of rubber tubing can be placed in the ripple tank to approximate reflection from a curved mirror. The curved mirror in Figure 16.4 yields a surprising result.

Straight waves traveling toward the curved barrier, as shown, are reflected as circular waves that come to focus at a point.

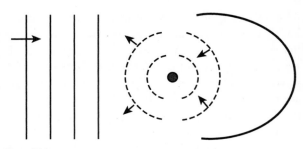

Figure 16.4

Straight waves reflect from a barrier shaped as a curve.

Refraction

The depth of water in the ripple tank can be changed so that there are extreme differences in depth by placing a glass or plastic plate just under the surface of the water. That was done for the diagram in Figure 16.5. The dark lines represent crests of straight waves moving from left to right in the diagram as indicated.

The line across the tank represents the change in depth of water from deep to shallow, reading left to right. The waves are periodic waves so both the incident wave in the deep water and the waves in the shallow water have the same frequency. Different depths of water are different media. You can see that the wavelength in the deep water is greater than the wavelength in the shallow water, so: $\lambda_d > \lambda_s$, $f\lambda_d > f\lambda_s$, and $v_d > v_s$.

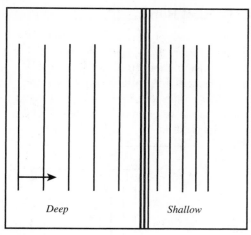

Deep   Shallow

Figure 16.5

When straight waves change media, from deep to shallow water, the wavelength changes.

The waves in Figure 16.6 cross a barrier at an angle. The water is deep on one side of the barrier and shallow on the other. The diagram in Figure 16.6 shows clearly that there is a change in direction as well as a change in speed. Changing the angle of the barrier for several trials reveals that Snell's law describes the change in direction and more. Look at the blown-up version of one incident wave and its corresponding refracted wave in the lower-left portion of Figure 16.6.

Physical Harm

The smaller of the two angles between the incident wave and the junction is the angle measured as the angle of incidence. The smaller of the two angles between the refracted wave and the junction is the angle measured as the angle of refraction.

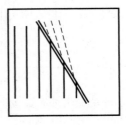

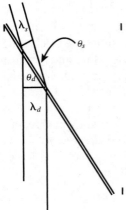

Figure 16.6

Straight waves change direction when they cross a barrier at an angle traveling from deep water to shallow.

A line segment is drawn representing the wavelength of each wave along with a label. Two right triangles are formed; when that is done, and the triangles share one hypotenuse, call it H. An expression for Snell's law may be written using the two right triangles: $\dfrac{\sin \theta_d}{\sin \theta_s} = \dfrac{v_d}{v_s}$.

Phun Phacts

The wave model correctly predicts the change in speed when the wave changes media. The change in direction is correctly explained with Snell's law.

That means that the speed in the second medium is less than the speed in the first medium when the angle in the second medium is less than the angle in the first medium. Water waves crossing a junction between media change direction and have a correct change in speed when considering waves as a model for light.

Diffraction

Water waves also bend around corners. That is called *diffraction*. Straight waves are generated in the ripple tank as shown in Figure 16.7. A barrier with a slit smaller than the wavelength of the wave is placed in the path of the straight waves.

The waves pass through the opening and curve, bending around the edges of the slit. The slit acts very much like a point source of waves. Remember when dipping the finger into the water created waves on the surface of the water?

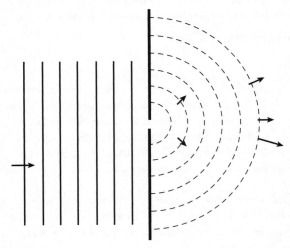

Figure 16.7

Straight waves passing through a slit bend sharply if the slit is small compared to the length of the wave.

The slit acts very much the same way if it is small compared to the length of the wave. If the slit is large compared to the length of the wave, then the wave passes through the slit and has only slight curving at its edges. In fact if the wavelength is very short compared to the width of the slit, the wave goes practically straight through the slit with very little curving.

Interference

The standing wave discussed earlier was generated by a source and its image through a junction. The image had to be creating waves with the same frequency and at the same time as the source. Two sources of periodic motion that are creating waves at the same time are said to be in *phase*.

The standing wave pattern had regions where crest met crest and trough met trough, creating a deeper trough and a higher crest that are called loops. Waves interacting in that way are said to *interfere constructively*. The nodes in the standing wave are regions of *destructive interference*.

The ripple tank is equipped with two plastic beads mounted on a rocking arm that causes the beads to dip into the water in phase. The rocking arm is driven by a small electric motor that has a frequency that is the frequency of the dipping plastic beads. The frequency of the motor is measured with a calibrated strobe light. While the frequency is being measured, the wavelength of the water waves generated by the dipping beads is determined at the same time.

As the waves from the two point sources travel through each other, they create regions of higher waves and regions of perfectly still water. The total picture is called an interference pattern. Lines of still water and lines of maximum disturbance, deeper troughs and higher crests, characterize the interference pattern. The lines of still water are called *nodal lines*. The nodal lines are the regions where a crest from one point source meets a trough from the other.

Regions of maximum disturbance occur where a crest from one source meets a crest from the other source and where a trough meets a trough. The drawing in Figure 16.8 helps to visualize the interference pattern of two point sources, in phase and with the same frequency, in two dimensions.

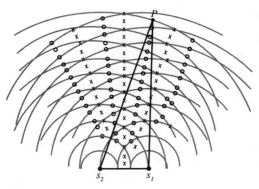

Figure 16.8

The nodal lines separated by regions of maximum disturbance of the medium characterize the two-dimensional interference pattern.

An Interference Pattern of Light

Young gave us the tool needed to view an interference pattern for light as well as the interference pattern in the ripple tank. Remember that two point sources in phase with the same frequency are needed to set up an interference pattern. In Young's experiment for light, one light source is used but it is viewed through two tiny slits very close together.

The two slits act like two point sources just as we found in the diffraction discussion. The two point sources are in phase and have the same frequency because the same wave from a source passes through both slits at the same instant.

Physical Harm

Young's experiment gave you the idea of two tiny slits a small distance apart diffracting a sample of the same wave from the same source to create an interference pattern. The slits act like two point sources in phase with the same frequency.

Suppose you take the pattern in Figure 16.8 and rotate the pattern about the line segment joining s_1 and s_2. Can you imagine nodal planes being generated by the nodal lines? If you view the nodal planes of an interference pattern of light, the nodal planes intersect the retina of your eye in lines and you see the lines of intersection. The diagram in Figure 16.9 helps to visualize the observed interference pattern and the experimental arrangement for viewing the nodal lines.

One side of a microscope slide can be painted with a quick-drying graphite suspension. When the graphite is dry, two double-edged razor blades can be held together and two slits can be scraped in the graphite with the edges of the blades. A thin line of graphite separates the slits by a distance d.

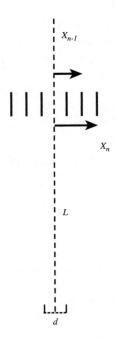

Figure 16.9

The nodal lines in the interference pattern for light are seen as black vertical lines separated by bright regions.

The distance *d* is measured with a micrometer caliper accurate to 0.01 mm. The source of light is placed about 2.0 m away against a wall. The light source is viewed through the tiny slits by holding the microscope slide close to the eye. The interference pattern in Figure 16.9 is observed having up to 10 nodal lines.

Notice that the nodal lines are the same distance apart. Even though x cannot be measured directly, calculate Δx, the distance between adjacent nodal lines. Let x_n represent the distance to the nth nodal line and x_{n-1} represent the distance the nodal line next to it. Then

$$\Delta x = x_n - x_{n-1}$$

$$\Delta x = \frac{L\lambda\left(n - \frac{1}{2}\right)}{d} - \frac{L\lambda\left((n-1) - \frac{1}{2}\right)}{d} \text{ using}$$

Young's Experiment.

$$\Delta x = \frac{L\lambda}{d} \text{ by simplification.}$$

The red light is found to have a wavelength of about 600 nanometers and the wavelength of blue light is about 400 nanometers. The wave model for light explains everything that the particle model does at this point and it does a better job explaining refraction and diffraction than the particle model does.

The Least You Need to Know

♦ A pulse from a large spring entering a small spring is partially reflected upside down at the junction.

♦ When a wave in a ripple tank crosses from deep into shallow water, it changes direction and slows down.

♦ Two point sources must be in phase and have the same frequency in order to generate an interference pattern.

Glossary

acceleration   The rate of change of velocity or the change in velocity divided by the time for that change to take place.

acoustics   The sound-producing qualities of a room or auditorium.

actual mechanical advantage   The gain a user realizes from a machine because it includes the effects of friction. Ideal mechanical advantage is the theoretical gain a user expects from a machine because it does not include the effects of friction.

adhesive forces   Forces of attraction between particles of different kinds of matter.

alpha particles ($_2^4He$)   Helium nuclei that have a positive charge and are emitted from radioactive elements like radium.

alternating current (AC)   An electric current that travels in one direction and then the opposite direction with a fixed period. Your local power company, through outlets in your home, provides AC.

ampere   The practical unit of electric current. The ampere is 1 coulomb of charge passing a point in an electric circuit in one second.

amplitude   The maximum displacement of the object undergoing simple harmonic motion.

anode   The positive electrode of a cell or battery.

atom   The smallest particle of element that exhibits the chemical properties of the element.

atomic mass number   The number of nucleons the atom contains.

atomic number   The number of protons in the nucleus of an atom.

average speed The change in position of an object divided by the change in time or the distance an object travels divided by the time required to complete the trip.

battery A combination of cells. It is constructed to overcome the limitations of one cell. A battery provides a larger current, a larger potential difference, or both.

beta particles High-speed electrons emitted by radio-active elements and have a negative charge.

branch (of a circuit) A division of a parallel part of the circuit.

British thermal unit (BTU) The quantity of heat required to raise the temperature of 1 pound of water 1 Fahrenheit degree.

calorie The quantity of heat needed to raise the temperature of 1 gram of water 1 Celsius degree.

capacitor An electrical device used to store an electrical charge.

cathode The negative electrode of a cell or battery.

center of curvature The center of the sphere of which the mirror is a part.

centripetal acceleration The center-seeking acceleration resulting from uniform circular motion.

centripetal force The center-seeking force on an object at every instant that deflects the object into a circular path at a constant speed.

closed polygon method A way of adding vectors that requires that any number of vectors may be added together by joining them together foot to head until the sum is complete. The resultant vector is found by closing the polygon with a vector drawn from the foot of the first vector in the sum to the head of the last in that order.

cloud chamber   A transparent container filled with alcohol vapor suspended between electrically charged plates. While a radioactive source emits alpha and beta particles, a source of light is directed at the alcohol-filled chamber. Trails much like vapor trails of jet planes are formed along the path where the radioactive particles have traveled. Their trails are easily viewed by the light reflected from the droplets of vapor in the cloud formed.

cohesive forces   The attractive forces between particles of the same kind.

components (of a vector)   Those parts whose sum is the given vector.

compression   That part of a longitudinal wave where the particles of the medium are pushed closer together.

concave spherical mirror   A segment of a sphere with the inside surface polished to reflect light.

condensation   The process of a vapor or a gas changing to a liquid.

conductors   Materials that allow electrons to move freely throughout the material.

convex spherical mirror   A segment of a sphere with the outside surface polished to reflect light.

corona discharge (brush discharge)   A bluish glow of ionized gases formed at any sharp point of a conductor that is under the influence of high-intensity electrons.

coulomb (C)   The fundamental unit of charge in the MKS system of measurement. The MKS system is the practical system for the study of electricity.

covalent bonding   The combination of two atoms to form a molecule by sharing a pair of electrons.

crest   That portion of the graph of a traverse wave that lies above the time axis.

critical angle The angle of incidence in the substance that has an angle of refraction of 90° in a second substance.

cycle One complete trip for an object moving in a circular path at a constant speed as well as the corresponding trip of its projection on the diameter of the circular path.

defining equation A statement of a relationship between two units of measurement.

density The amount of matter in a unit volume. Since matter is measured in two different ways, there are two types of density. Mass density is the amount of mass in a unit volume of matter and weight density is the amount of weight in a unit volume of matter.

derived quantities Physical quantities that are defined using two or more fundamental quantities or one fundamental quantity used more than once.

destructive interference Waves interact to cancel each other out or destroy each other.

diffraction The spreading out of waves as they pass through an opening or pass by a sharp corner.

diffusion The movement of particles of one kind of matter into the empty space of a different kind of matter because of the random motion of the particles.

direct current (DC) An electric current that travels in only one direction in the circuit. It travels from the negative side of the cell, through the circuit external to the cell, to the positive side of the cell.

displacement A vector defined as a change in position.

Doppler effect The shift in the pitch of a source of sound because of the relative motion between the source and the observer.

dyne The force required to accelerate a mass of 1 gram at a rate of 1 cm/s^2.

efficiency   A fraction greater than zero and less than one that expresses what part of input work the output work amounts to. The efficiency is usually stated as a percent obtained by multiplying the fraction by 100 percent.

electric circuit   The conducting path for the flow of charge.

electric current   The flow of charge past a point in an electrical circuit for each unit of time.

electron   A particle that has a charge of $1.6 \times 10^{-19} C$ and a mass of $9.1 \times 10^{-31} kg$.

element   A substance like copper, hydrogen, or neon that cannot, by chemical means, be broken down into other substances.

equal vectors   Have equal magnitudes and the same direction.

equilibrium   When the sum of the forces is zero. If one force is used to balance the effects of two or more other forces, then that force is called the equilibrant and is equal in magnitude and opposite in direction to the resultant of the other forces.

evaporation   The process of a liquid changing to a gas or a vapor.

focal point (of a concave spherical mirror)   A point on the principal axis where all rays intersect after they are reflected from the mirror. That is true if they are rays that are parallel to the principal axis and near the principal axis. The focal point is halfway between V and C. That means the distance from V to C is $2f$ where f is the focal length of the mirror, the distance from V to F.

force of gravity   The downward pull of the earth on any mass placed above, on, or below the earth's surface. This is a pull at a distance because nothing is attached to the mass pulling it toward the earth.

freely falling body   An object that is influenced by the force of gravity alone.

freezing point   The temperature at which a liquid changes to a solid.

frequency   The reciprocal of the period and is given by $f = \dfrac{1}{T}$ and has units of $\dfrac{cycles}{s}$ or just $\dfrac{1}{s} = s^{-1}$ since a cycle is not a unit of measurement. Another common unit of measurement of frequency is the Hertz so $1Hz = s^{-1}$.

fundamental quantities   The building blocks for the foundation of physics. Time, space, and matter are the quantities required to study that area of physics called mechanics.

fusion   The process of changing from a solid to a liquid.

gamma rays   Very penetrating short wavelength electromagnetic radiation emitted from radioactive elements.

graphic solution   Achieved by using drawing instruments to construct a scale drawing. A ruler provides the measurement of magnitude to scale and the protractor enables you to measure angles.

gravitational field   That region of space where the force of gravity acts on a unit mass at all locations on or above the surface of the earth.

half-life   The time required for half of the atoms in a given sample of a radioactive element to decay.

heat   The kinetic energy of the particles of matter.

heat capacity   The quantity of heat required to raise the temperature of a body 1 degree.

heat of vaporization   The quantity of heat required to change a unit mass or weight of a liquid to a gas or vapor at the normal boiling point.

hydrometer   An instrument used to measure the specific gravity of a liquid.

illumination   The luminous intensity at a distance from a source. The unit of illumination is the foot-candle (ft-cd).

impulse   A physical quantity that results from a force being applied to a body for a certain amount of time.

inertial frame of reference   A frame of reference that is at rest or moving at a constant speed. For our purposes, the earth is an inertial frame of reference where the laws of Newton hold.

instantaneous velocity   The velocity of an object at any instant of its motion. The instantaneous velocity vector is tangent to the path of motion of the object at every point along the path.

insulator   A material that does not allow free movement of charge.

interfere   Waves that move through each other and either reinforce each other or destroy each other. Waves interfere constructively when they interact to build each other up or reinforce each other.

ion   A charged particle caused by an atom gaining or losing electrons to exhibit excess or a deficiency of electrons.

ionic bonding   The formation of a molecule by the transfer of an electron from one atom to the other creating two ions more stable than the atoms were before the transfer. The two newly formed ions attract each other, due to opposite electrical charge, forming a molecule.

junction   Where two media are joined together. That is, a spring can be anchored at a junction, two springs may be attached at a junction, or a spring may be attached to a string at a junction.

kilocalorie   The quantity of heat required to raise the temperature of 1 kilogram of water 1 Celsius degree.

kinetic energy   The energy an object has when it is in motion.

law of heat exchange   The heat gained by cold substances is equal to the heat lost by hot substances.

longitudinal wave   A disturbance traveling through a medium in which the particles vibrate in paths parallel to the direction the wave is traveling.

loop (or anti-node)   The region of reinforced amplitude in a standing wave pattern.

lumen   The unit of intensity of illumination.

luminous intensity   The strength of a source of light measured in a unit called the candela.

mass number   The whole number nearest the atomic mass of the element.

matter   Anything that occupies space and has mass.

maximum height (of a projectile)   The greatest distance the projectile is above the earth or the level of launch.

melting point   The temperature at which a solid changes to a liquid at standard pressure.

metallic bonding   The name of the attraction atoms of solid metals have for each other in a closely packed arrangement due to the continuous exchange of loosely held electrons in the outer energy levels of the atoms of the metal.

mole   Avogadro's number of items, 6.02×10^{23} items, or a basic unit of quantity.

molecule   The smallest particle of a compound.

momentum   A physical quantity that has its magnitude determined by its mass and velocity.

move   An object experiences a change in position.

neutron   A particle with no charge that has about the same mass as the proton.

newton   The force required to accelerate a mass of 1 kilogram at a rate of 1m/s^2.

nodal lines   The regions in an interference pattern where the waves from two point sources that are in phase and with the same frequency cancel each other out. The destructive interference takes place where a trough from one source meets a crest from the other source.

node (or nodal point)   The points in a standing wave pattern where the superposition of the waves produces zero amplitude. Nodes are half a wavelength apart as are the loops of a standing wave.

normal   A line perpendicular to a line or to a surface.

nucleus   The core of the atom that is made up of protons and neutrons.

parallelogram method   A way of adding vectors requires that the feet of two vectors to be located at the same point. A parallelogram is then constructed by drawing a line parallel to one of the vectors through the head of the other. A second line is constructed parallel to the second vector passing through the head of the first. The resultant vector is found by joining the feet with the opposite vertex of the parallelogram in that order.

period   The time for the object in simple harmonic motion to complete one cycle.

phase   The position and direction of movement of the object undergoing periodic motion.

pith   A very light dry fibrous material. Pith is the central column of spongy cellular tissue in the stems and branches of some large plants.

position   A vector used to locate a point from a reference point or frame of reference.

potential difference (between two points in an electric field)   The work done per unit charge as the charge moves between those points.

potential energy The energy an object has because of its placement in a force field.

pound The force required to accelerate a mass of 1 slug 1ft/s^2.

pressure A quantity determined by the force on a unit of area.

principal axis (of a concave spherical mirror) The line that passes through the center of curvature and intersects the mirror at a point called the vertex of the mirror.

principal axis (of a lens) The line passing through the principal focus and the center of the lens.

proof-plane A small metal disk attached to an insulator. The disk may be rubbed on the charged insulator to sample the charge and limit the number of charges taken. The physical size of the disk limits the size of the charge to be placed on the electroscope.

propagation The sending out or spreading out of the wave from a source.

proportion An equation each of whose members is a ratio.

proton A positively charged particle with a mass of about $1.7 \times 10^{-27} kg$.

pulse A wave or a disturbance of short duration.

quantitative description (of images) Involves actual measurements of distances such as f, S_o, and S_i.

range (of a projectile) The maximum distance traveled horizontally by the projectile. It is measured from the point of launch to the point of return to the same level.

rarefaction That part of a longitudinal wave where the particles of the medium are being spread apart.

reflection The change in direction of light when it strikes a smooth surface causing the light to bounce off the surface.

refraction   The change in direction of light when it leaves one medium and enters a different medium.

regelation   The melting of ice under pressure and then freezing again after the pressure is released.

resistance (in an electric circuit)   The opposition to the flow of charge or current in the circuit.

resistor   A component in an electrical circuit used to establish the amount of current and/or potential difference at different places in the circuit.

resolution   The name of the process of identifying the parts of a vector.

resonance (in sound)   The increased amplitude of vibration of an object caused by a source of sound that has the same natural frequency.

resultant vector   The name assigned to the vector representing the sum of two or more vectors.

ripple tank   A laboratory instrument used to study waves in two dimensions. It is a shallow tank that holds water about 1 to 2 centimeters deep. It has a glass or clear plastic bottom about 24 inches square.

scalar   A quantity that has magnitude only. Any units of measurement are included when we refer to magnitude.

scalar multiplication   The product of a scalar and a vector that results in a vector with a magnitude determined by the scalar. The direction of the product is the same as the original vector if the scalar is positive, and opposite the direction of the original vector if the scalar is negative.

schematic diagram   A symbolic representation of electric circuits. That is, symbols representing components, conductors, and instruments of measurement are used instead of pictures of those items.

significant figures Those digits an experimenter records that he or she is sure of plus one very last digit that is doubtful.

simple harmonic motion The to and fro motion caused by a restoring force that is directly proportional to the magnitude of the displacement and has a direction opposite the displacement. An example is the motion exhibited by a vibrating string or a simple pendulum.

simple pendulum A physical object made up of a mass suspended by a string, rope, or cable from a fixed support. It is called a simple pendulum because the string, rope, or cable has a negligible amount of mass compared to the mass of the object being supported. A physical pendulum is a physical object that has the mass distributed along the full length of the pendulum, and the simple pendulum does not have most of the mass concentrated in one place.

solidification The process of changing from a liquid to a solid.

specific gravity The ratio of the weight density of the liquid to the weight density of water. It is also the ratio of the mass density of the liquid to the mass density of water.

specific heat The ratio of its heat capacity to the mass or weight of a substance.

spring balance A device containing a spring that stretches when pulled by a force. The amount of stretch is calibrated to correspond to force units, nt, dynes, or lb.

standing wave The superposition of two waves of the same frequency moving in different directions in the same medium.

sublimation The direct change of a solid to a vapor without going through the liquid state.

superpose   When one wave adds onto the other. During the instant of superposition, neither pulse is recognizable individually because they appear as one combination of pulses.

surface tension   A quantity or condition of the surface of a liquid that causes it to tend to contract.

temperature   The condition of a body to take on thermal energy or to give up thermal energy.

terminal velocity   The maximum velocity of a falling object in air.

thermodynamics   The study of quantitative relationships between other forms of energy and thermal energy.

time of flight (of a projectile)   The total time it is in the air.

trajectory   The arced or curved path of a projectile.

transverse wave   A disturbance traveling through a medium in which particles of the medium vibrate in paths that are perpendicular to the direction of propagation of the wave.

trigonometry   A branch of mathematics that deals with relationships of angles and corresponding sides of triangles.

triple beam balance   A device for measuring mass by comparing an unknown mass with a known mass by balancing two pans. The two pans are attached to a beam that is supported by a fulcrum in much the same way as a seesaw or teeter-totter.

trough   That portion of the graph of a transverse wave that lies below the time axis.

unified atomic mass unit (u)   The unit of mass for used to stipulate nuclear masses.

uniform motion   Motion characterized by a constant speed.

unit analysis The process of defining a disguise of unity (one) in terms of units of measurement that will enable you to change from one unit of measure to a larger or smaller unit of measure without changing the value of the measured quantity.

valence electron An electron in an incomplete shell that an atom can lose to become stable or for a different atom to gain to become stable.

Van de Graaff generator A motorized source of high concentration of electrons with high potential energy.

vector A quantity that has magnitude, direction, and obeys a law of combination.

velocity The rate of change of displacement.

voltmeter The instrument used to measure potential difference in an electrical circuit.

wavelength (of a transverse wave) The distance from the beginning of a crest to the end of an adjacent trough. One wave is made up of a crest and a trough. The distance from the beginning of a crest to the end of the same crest is one half of a wavelength.

weight The force of gravity of the earth on any object. It is a vector quantity with direction radially inward toward the center of the earth. The direction as described locally is down.

Index

A

accelerated motion,
  uniform, 19
acceleration, 18–19, 32
  centripetal, 34–35
  due to a force acting at
    a distance, 51
  graphical techniques,
    38–43
addition
  rounding, 8
  vectors, 24–27
adhesive forces, 82–86
air friction, 49–51
alpha particles, 138
ammeter, 166
atomic mass number, 135
atomic nucleus, 138
atomic number, 135
atoms, 132–133
Avogadro's Number, 7

B

Bohr, electrons and, 132
boiling, 112
boiling point of water,
  106

Boyle's law, 90, 92
buoyant force, liquids and
  pressure, 96, 99, 102

C

calculations, significant
  figures and, 10, 12
calories, 108–109
Celsius scale, 106
centripetal acceleration,
  34–35
  deflecting force and,
    59
CGS (centimeter, gram,
  second), 2
change, 14
charged rain clouds, 157
circular motion, 33–34
cohesive forces, 82, 84–86
components of vector, 30
condensation, 112
conduction, electrical
  charge, 150, 153
conductors
  electrical charge and,
    146–147
  electric circuit, 164

constructive interference, 205

converging lens, refraction of light and, 188

convex lens, refraction of light and, 188

coulombs, 148

covalent bonding, 141

critical angle for glass, refraction and, 184

current electricity, electrons and, 159

D

decimal point, zeros and, 4

defining equations, 4

deflecting force, 58

density, 82
    gases, 89

destructive interference, 205

diffraction, 203–204

diffusion, 82

displacement, 31

distance from Earth to sun, 7

division, significant figures, 8

Doppler effect, 127

E

Earth, distance to sun, 7

electrical charge
    charged rain clouds, 157
    conduction, 150, 153
    conductors and, 146
    induction, 150, 153, 155
    insulators and, 145
    insulators and charging, 146, 148
    lightning, 157
    Van de Graaff generator, 155

electric circuit, 164–169, 173

electric current, 161–164

electron transfer, 140

electron exchange, 140

electrons
    Bohr and, 132
    current electricity and, 159
    metals and, 141
    negative charge, 132
    Rutherford and, 132
    Thomson, 132

electron sharing, 140

energy
    ground, 147
    kinetic energy, 71
    potential energy, 75

thermal energy
  calculating, 109
  zero potential, 147

F

Fahrenheit scale, 106
falling object, motion, 52
final position, 14
final time, 14
force
  acceleration and, 51
  air friction, 49–50
  components of, 45
  deflecting, 58
  due to gravity, 52, 54
  graphical techniques,
    38–43
  power and, 69
  pressure and, 86
  pressure and buoyant
    force, 96, 99, 102
  pressure and liquids,
    96
  uniform motion and,
    46–47
  vector nature of, 43–44
FPS (foot, pound, sec-
  ond), 2
freely falling bodies,
  50–52
freezing point, 111
freezing point of water,
  106

frequency, sound, 124
fusion, 110, 113

G–H

gases, 87, 89
graphic solution, 44
gravitational mass, inertial
  mass and, 53
gravity
  force due to, 52, 54
  universal law of gravi-
    tation, 61
ground energy, 147

harmonic motion, 59, 61
heat, 104
  calories, 108–109
  fusion and, 110, 113
  heat capacity, 109
  specific heat, 109
  vaporization and, 110,
    113
  work and, 109, 113
Hook's law, 83
horizontal motion, 55
hydrogen atoms, 132
hydrogen molecules, 131

I

impulse, 78–79

induction, electrical
  charge, 150, 153, 155
inertial balance, 53
inertial mass, 53
initial position, 14
initial speed, 14
initial time, 14
insulators, electrical
  charge and, 145
    charging, 146, 148
interference, 205–206,
  209
interference of sound, 128
inverse square law, light,
  176–177
ionic bonding, 140
ions, 136–139
isotopes, 135–136

K

Kelvin scale, 106
kinetic energy, 71
    projectiles, 74
    work and, 71, 74
Kirchhoff's first law, 168

L

Leyden jar, 160
light
    beam of, 177
    direction, 177
    inverse square law,
      176–177

particle model, 175
particles, rays, 177
pencil of, 177
reflection, quantita-
  tive description of an
  image, 181
reflection of, 178–179
refraction, 182, 184,
  188, 190
speed, 177
lightning, 157
lightning rods, 161
liquids, 84
    adhesive forces, 84, 86
    cohesive forces, 84, 86
    Pascal's principle, 94
    pressure, 86, 93–96,
      99, 102
longitudinal waves, 119

M

matter, 2
measurement
    matter, 2
    recording, 3
    significant figures, 5
    space, 2
    time, 2
    unified atomic mass
      unit, 135
melting point, 110
metals, electrons and, 141
MKS (meter, kilogram,
  second), 2

mole, 7
molecules
    covalent bonding, 141
    hydrogen, 131
    sugar, 131
    water, 131
momentum, 78–79
motion, 14
    acceleration, 18–19
    average velocity, 14
    change, 14
    circular, 33–34
    falling objects, 52
    final position, 14
    final time, 14
    final velocity, 14
    horizontal, 55
    initial position and, 14
    initial speed, 14
    initial time, 14
    projectile, 54–58
    simple harmonic
        motion, 59, 61
    sound and, 124
    uniform, 15, 18, 46–47
    uniformly accelerated,
        19
    vertical, 56
multiplication
    scalar, 28
    significant figures, 8
    vectors, 28

N

nanometers, 132
negative charge, electrons
    and, 132
neutrons, atoms, 133
Newton, universal law of
    gravitation, 61
Newton's second law, 43
nonzero digits, 4
nucleons, atoms, 133

P

parallel circuits, 168–169,
    173
particle model of light,
    175
Pascal's principle, 94
position, 31
    velocity and, 31
potential difference,
    163–164
potential energy, 75, 77
power, 69
pressure
    force and, 86, 96, 99,
        102
    liquids, 86
    liquids and, 93–96, 99,
        102
principal axis, refraction
    of light, 190

projectile motion, 54–58
projectiles, kinetic energy, 74
protons, atoms, 133
  atomic number, 135

Q–R

quantitative description of an image, 181

recording observations, 3
reflection, waves
  at junction of two media, 194
  curved mirror, 199
  speed, 196
  transversed, superposed, 195–196
  two-dimensional, 198
reflection of light, 178–179, 181
reflection of sound, 127
refraction, 200, 203
refraction of light, 182
  converging lens, 188
  convex lens, 188
  critical angle for glass, 184
  principal axis, 190
  prisms, 184
  Snell's Law, 182, 184
reporting observations, 3
resistance, 164
resistance, electric circuit, 164

resolution, vectors, 30
rounding, 8
Rutherford
  atomic nucleus, 138
  electrons and, 132

S

scalar multiplication, 28
scalars, 23
scientific notation, 5–6
  Avogadro's Number, 7
series-parallel circuits, 169, 173
series circuit, electric circuit, 165, 167
SI (International System of Units), 2
significant figures, 3, 5
  calculations and, 10, 12
simple harmonic motion, 59, 61
Snell's Law, 182
  two media other than air, 184
solidification, 111
solids, 81-82
sound
  Doppler effect, 127
  frequency, 124
  interference, 128
  motion and, 124
  reflection of, 127
  source, 115–118
  speed, 123–124

temperature and, 126
soundwaves, 118
    compression, 122
    crest, 121
    longitudinal, 119
    properties, 120, 123
    rarefaction, 122
    transverse, 119
    trough, 121
    wavelength, 121, 124
space, 2
specific heat, 109
speed, 13
    acceleration, 18
    acceleration and, 32
    average velocity, 14
    centripetal accelera-
        tion, 34–35
    circular motion, 33–34
    final velocity, 14
    initial, 14
    light, 177
    sound, 123–124
    waves, reflection, 196
static charge, 143
    detectors, 148, 150
    negative, 144
    positive, 144
subtraction
    rounding, 8
    vectors, 27
sugar molecules, 131
sun, distance from Earth,
    7

T

temperature, 104
    boiling point of water,
        106
    Celsius scale, 106
    change, 107
    change in, 105–106
    Fahrenheit scale, 106
    freezing point, 111
    freezing point of water,
        106
    heat, 104
    Kelvin scale, 106
    melting point, 110
    sound and, 126
    thermometers, 104
terminal velocity, 51
thermal energy, calculat-
    ing, 109
thermometers, tempera-
    ture, 104
Thomson, electrons and,
    132
Three, 24
time, 2
    final time, 14
    initial time, 14
transversed waves, super-
    posed, 195–196
transverse waves, 119
triple beam balance, 53
truth, 8, 9
two-dimensional waves,
    198

U

unified atomic mass unit, 135
uniformly accelerated motion, describing, 19
uniform motion
    describing, 15, 18
    net force and, 46–47
unit analysis, 5
universal law of gravitation, 61

V

Van de Graaff generator, 155
vaporization, 110, 113
vectors, 24
    addition, 24–27
    components, 30
    description, 23
    force and, 43–44
    multiplication, 28
    resolution, 30
    subtraction, 27
velocity
    acceleration, 18
    acceleration and, 32
    centripetal acceleration, 34–35
    circular motion, 33–34
    displacement and, 31
    position and, 31
    terminal velocity, 51
vertical motion, 56

W–X

water molecules, 131
wavelength, sound, 121, 124
waves
    at junction of two media, 194
    curved mirror, 199
    reflection speed, 196
    two-dimensional, 198
weight, force due to gravity, 52–54
work, 63
    definition, 64
    displacement and, 64
    heat and, 109, 113
    kinetic energy and, 71, 74
    potential energy and, 75, 77
    power and, 69

Y–Z

zero potential energy, 147
zeros, 4
    decimal point and, 4